OUTCOMES EDITION

for Papua New Guinea

Mathematics

8B

Sue Gunningham

Pat Lilburn

OXFORD
UNIVERSITY PRESS

Level 8, 737 Bourke Street, Docklands, Victoria 3008, Australia.

Oxford University Press is a department of the University of Oxford. It furthers the University's objective of excellence in research, scholarship, and education by publishing worldwide in

Oxford New York

Auckland Cape Town Dar es Salaam Hong Kong Karachi
Kuala Lumpur Madrid Melbourne Mexico City Nairobi
New Delhi Shanghai Taipei Toronto

With offices in

Argentina Austria Brazil Chile Czech Republic France Greece Guatemala
Hungary Italy Japan Poland Portugal Singapore South Korea Switzerland
Thailand Turkey Ukraine Vietnam

First published 2008
Reprinted 2009 (twice), 2010, 2011, 2013, 2014, 2021, 2023, 2026

ISBN 978 0 19 556251 4

Typeset by Palmer Higgs
Illustrated by Uramina and Nelson Pty Ltd and Guy Holt
Printed in Singapore by Markono Print Media Pte Ltd

Table of Contents

For Students

Dear Student,

We hope you enjoy working through the Oxford Mathematics books for Grade 8. The two books have been written to show how mathematics is useful in dealing with the everyday world. The lessons will help you to gain important life skills that will be useful now and in the future when you leave school to work for someone else or to set up your own small business.

The books comprise four topics:

Book A **Topic 1:** Gadgets

Topic 2: And the Winner is . . .

Book B **Topic 3:** A Lot Like Me

Topic 4: The Tourism Market

Each topic is broken into three *Learning Units* that will build on your previous knowledge and introduce you to new concepts. Each Learning Unit begins with pictures and open-ended questions to explore your knowledge of both the context and some of the mathematics to be covered.

Help Boxes have been located throughout the Learning Units to provide clear instructions to support your understanding of particular concepts. We encourage you to refer back to these Help Boxes as you progress through the book to refresh your memory and consolidate your understanding. A number of *Challenges* also appear in the Learning Units. It is hoped that you will attempt as many of these Challenges as possible to extend your thinking.

A Revision Unit, titled *Additional Learning, Revision and Assessment*, appears at the end of each topic after the three Learning Units. This unit has four parts. It contains one stand-alone lesson with a specific maths focus that is separate from those concepts contained in the three Learning Units. The stand-alone lesson is followed by a revision exercise that allows you to revise your understanding of the work covered in the topic. The unit also provides one investigative group task and a formal test.

A *Glossary* appears at the back of each book. The words in the glossary relate specifically to the content of that particular book and the information provided should make it easier for you to understand the meaning of different mathematical terms.

An answer section has been included at the back of each book.

We hope you find the books an interesting and challenging part of your learning journey.

The Authors

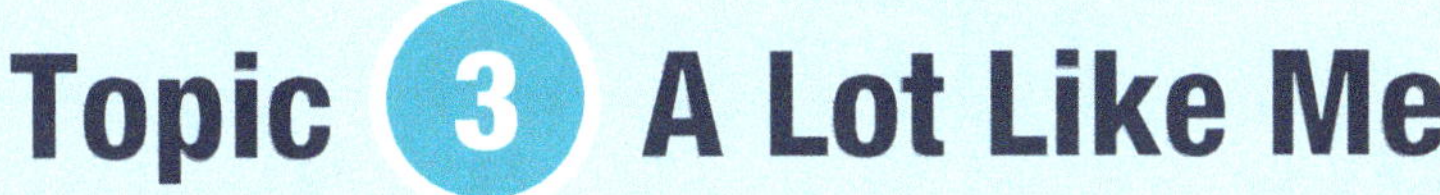

Topic 3 A Lot Like Me

Learning Unit 1: Different People, Different Sizes

Learning Unit 2: The Past and the Future

Learning Unit 3: Growing and Changing

Learning Unit 4: Additional Learning, Revision and Assessment

Topic 3 A Lot Like Me

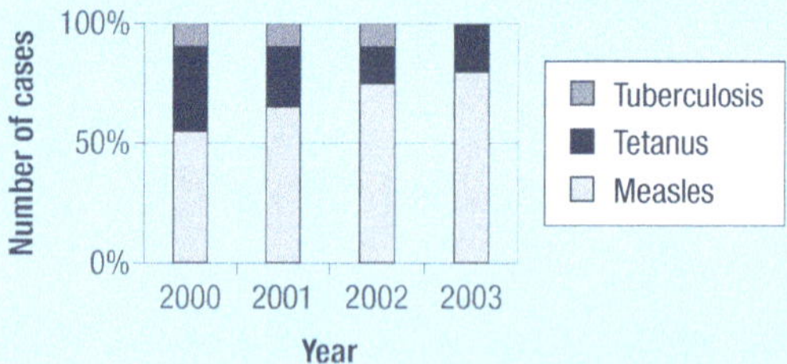

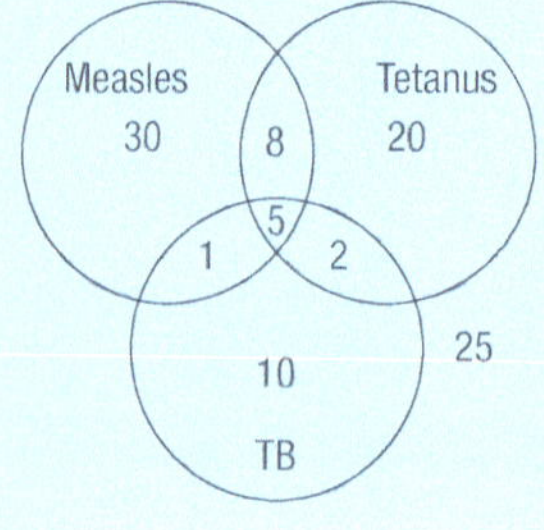

In this topic you will look at some of the health issues facing Papua New Guinea, as well as some of the programs put in place to reduce the spread of infectious diseases.

You will investigate different ways of representing and analysing statistical information, and the potential of making decisions about the future on the basis of trends identified in data.

The Assessment Tasks provide opportunity for you to demonstrate your understanding of the material covered both through project work and a formal test.

Topic 3: *A Lot Like Me* comprises the following Learning Units:

Learning Unit 1: Different People, Different Sizes

Learning Unit 2: The Past and the Future

Learning Unit 3: Growing and Changing

Learning Unit 4: Additional Learning, Revision and Assessment

Overview

Each Learning Unit reflects an aspect of mathematics used in everyday life. A brief summary of the mathematics in each Learning Unit appears below.

Learning Unit 1, *Different People, Different Sizes*, focuses on using weight and height to calculate a person's body mass. Body mass can be used to establish if a person fits within a 'healthy range' for people of a similar age.

Learning Unit 2, *The Past and the Future*, explores various ways of presenting data and calculating the frequency of certain data within a set. The benefits and limitations of using a random sample are also discussed in this unit.

The work in **Learning Unit 3**, *Growing and Changing*, investigates factors of numbers and the use of index notation as an efficient method for writing large numbers. The unit also builds on previous work dealing with fractions, decimals and percentages.

Learning Unit 4, *Additional Learning, Revision and Assessment*, includes a stand-alone lesson about scientific notation, a revision lesson, an investigation and a formal test.

Some of the materials that you may need to complete the activities in this topic are listed below.

You may need:

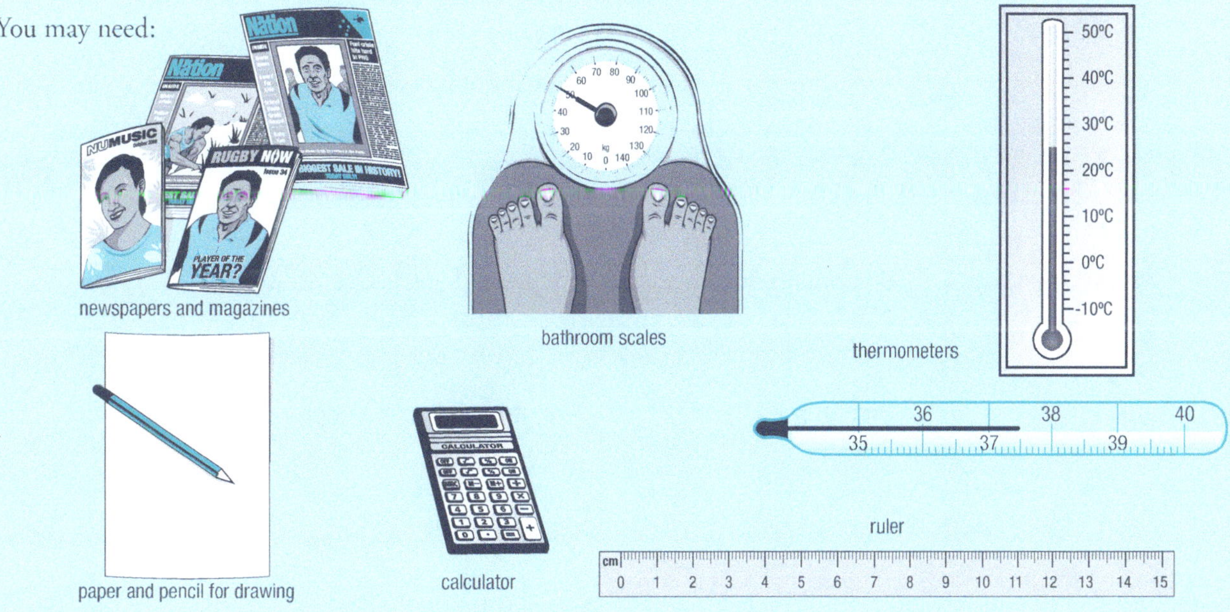

newspapers and magazines

bathroom scales

thermometers

ruler

paper and pencil for drawing

calculator

Learning Unit 1 Different People, Different Sizes

Strand: Number and Application

Fractions	Outcome 8.1.1	Apply fractions in problem solving
Ratios and Rates	Outcome 8.1.5	Apply ratios in solving problems from real life

Strand: Measurement

Weight	Outcome 8.3.1	Solve real life weight problems with confidence and competence
Weight	Outcome 8.3.2	Differentiate between weight and mass
Temperature	Outcome 8.3.3	Display temperatures including those from specialist thermometers

Strand: Chance and Data

Accuracy and Error	Outcome 8.4.5	Represent levels of accuracy
Estimation	Outcome 8.4.6	Identify and select appropriate estimation strategies

Strand: Patterns and Algebra

Algebra	Outcome 8.5.3	Manipulate simple algebraic expressions and solve real life problems

Lesson 1: Introduction	Promoting healthy lifestyles
Lesson 2: Weighing in	Comparing mass and weight
Lesson 3: Weight for age	Analysing weight for age charts
Lesson 4: Growing heavier	Solving problems involving weight Using ratio in real life situations
Lesson 5: Growing taller	Solving problems about growth over time Multiplying fractions mentally Using simple algebraic formulae
Lesson 6: Body Mass Index	Calculating Body Mass Index Interpreting data related to body mass index
Lesson 7: Pulse rate and percentage error	Deciding levels of accuracy Calculating percentage error
Lesson 8: Estimating measurement	Identifying strategies to minimise estimation errors
Lesson 9: How hot is it?	Reading temperature Calculating relative humidity

Lesson 1 Introduction

In this unit you will examine some of the mathematics involved in deciding the health of an individual. The relationship between age, height and weight will be used to determine a person's health. Body temperature is also an indicator of health, and the unit will look at reading thermometers and calculating changes in temperature.

There are many things that we can do ourselves to ensure our own health and well-being. Generally people concentrate on improving their fitness level, watching their diet and minimising their exposure to infections and high-risk situations.

Here are some slogans created by different health organisations to promote healthier lifestyles.

TAKE A WEIGHT OFF YOUR MIND

MOVE IT OR LOSE IT

EAT WELL, LIVE WELL

GO FOR YOUR LIFE

GET FIT, BE A HIT

EAT A LITTLE, WALK A LOT

1. Write a few sentences on each slogan to explain the message that the promoters are trying to give us.
2. List the things you do each day to ensure your physical health.
3. Which of these slogans would be of most interest to you? Why?
4. Consider the lifestyles, activities and food intake typical of many of the students at your school. Identify an issue of concern and create a catchy slogan to target that issue.
5. Make the slogan into a small banner and display it around the school.

Lesson 2 Weighing in

The terms 'weight' and 'mass' are commonly used to mean the same thing. However, in reality, 'weight' is a measurement of the pull of gravity on an object, whereas 'mass' is a measure of the amount of matter something contains.

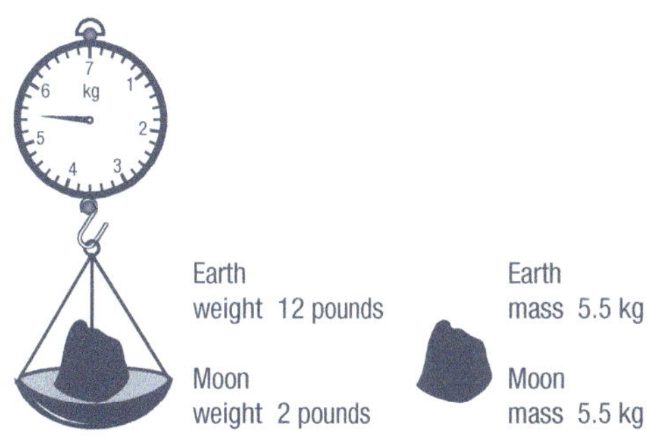

For example, an object may weigh 60 kilograms on Earth. The same object would weigh only 10 kilograms on the moon (due to the lower gravitational pull on the moon). The mass of the object, however, would be 60 kilograms on Earth and 60 kilograms on the moon because the amount of matter is unchanged. In other words, weight (how heavy an object is) is less on the moon but mass (the amount of matter) does not change.

For everyday use the difference between mass and weight is irrelevant, because everyone is on the Earth. However for calculations, such as for space launches, the difference is important. Scientists calculate knowing that the statement 'an object with a 1 kilogram mass also has a weight of 1 kilogram' is true on Earth but not elsewhere.

1 Venus has a gravitational pull that is about 0.88 of the Earth's.

- a If an object weighs 110 kg on Earth, what will it weigh on Venus?
- b If an object has a mass of 50 kg on Earth, what will be its mass on Venus?

2 Jupiter has a gravitational pull that is about 2.34 of the Earth's.

- a If an object weighs 165 kg on Earth, what will it weigh on Jupiter?
- b If an object has a mass of 75 kg on Earth, what will be its mass on Jupiter?

3 The moon has a gravitational pull that is about 0.165 of the Earth's.

- a If an object weighs 200 kg on Earth, what will it weigh on the moon?
- b If an object has a mass of 90.9 kg on Earth, what will be its mass on the moon?

4 Mars has a gravitational pull that is about 0.38 of the Earth's.

- a If an object weighs 185 kg on Earth, what will it weigh on Mars?
- b If an object has a mass of 84.09 kg on Earth, what will be its mass on Mars?

5 Neptune has a gravitational pull that is about 1.125 of the Earth's.

a If an object weighs 200 kg on Earth, what will it weigh on Neptune?

b If an object has a mass of 90.9 kg on Earth, what will be its mass on Neptune?

6 a Write your own mass in kilograms.

b Calculate what you would weigh in kilograms on the moon with a gravitational pull of 0.165 of the Earth's.

7 Copy the table into your workbook. Complete the table by entering your mass and then by using the information on gravitational pull to calculate your weight on each planet.

	My mass in kilograms		
	My weight on Earth in pounds (mass × 2.2)		
	Planet	**Gravitational pull compared to Earth**	**My weight on that planet**
a	Mercury	0.284	
b	Venus	0.88	
c	Mars	0.38	
d	Jupiter	2.34	
e	Saturn	0.925	
f	Uranus	0.795	
g	Neptune	1.125	

8 Order the planets from the one on which you weigh the most to the one on which you are lightest.

Challenge

Astronauts from the United States collected rocks that weighed 2 pounds, 7 pounds and 13 pounds on the moon. Pounds are an imperial measurement. The conversion from pounds to kilograms can be estimated using the formula:

$$\text{kilograms} = \frac{\text{weight in pounds}}{2.2}$$

Use this formula to help calculate:

a the total weight of the rocks on Earth

b the total mass of the rocks in kg

Lesson 3 Weight for age

All children grow at their own pace. This means there is a wide variation in the height and weight of 'normal' children. Many factors, including genes and diet, influence a child's growth and development.

Doctors and nurses use growth charts to decide if a child's height and weight is within the 'normal' range for their age. Growth charts represent data collected from thousands of children. The average height and weight for each age are established and plotted to form a curved line that became the 50th percentile curve. This means that in 1000 children, 500 would be above the line and 500 would be below.

Look at the growth chart for girls from birth to 36 months.

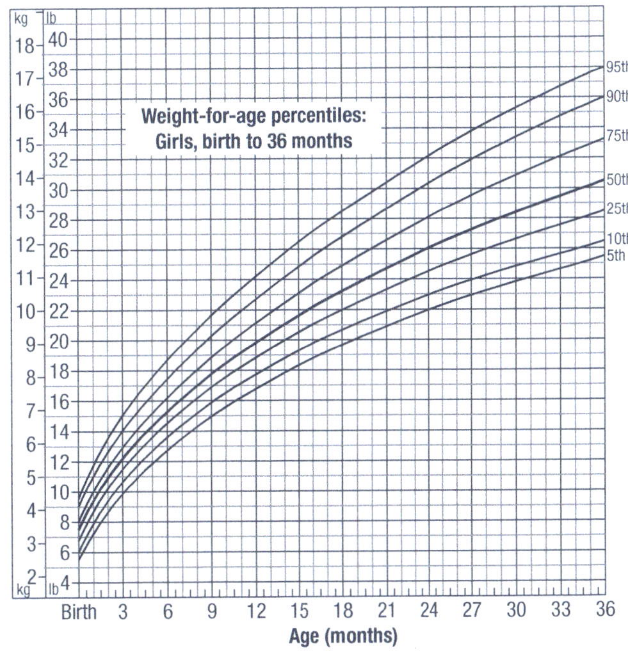

1 Answer these questions using data from the 50th percentile curve.

- a What is the average female birth weight in kilograms?
- b About how old is the average girl when she weighs 9 kg?
- c About how many kilograms does the average girl gain from birth to 12 months?
- d How does this compare with her weight gain for the next 12 months? Show your working.
- e The line begins very steeply and then gradually becomes less steep. Explain what this tells us about the average baby's growth pattern.

Challenge

During which three-month period from birth to 3 years does the average girl have the greatest weight gain? How do you know this?

Look at the growth chart on the right for boys from birth to 36 months.

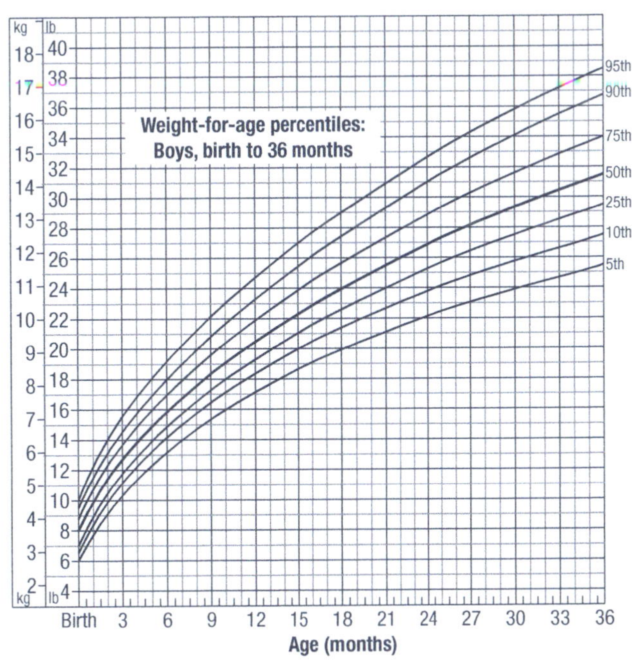

2 Answer these questions using the 50th percentile curve for both the boys' and the girls' charts.

a What is the average male birth weight in kilograms?

b What is the total weight gain (in kg) by the average boy from birth to 36 months?

c What is the difference in kilograms between the average birth weight of each sex?

d Describe the difference between the average boys' weight gain from birth to 15 months and from 15 to 36 months.

e About how much does the average boy weigh in kilograms at 36 months?

f Write a few sentences to describe some of the main differences you notice between the boys' and girls' data.

While the 50th percentile line plots the average child, the two graphs also show the curved line representing the 10th, 25th, 75th, and 90th percentiles. Ten per cent (100) of one thousand normal children will be at the 10th percentile. In other words, a boy whose weight was on the 10th percentile would be heavier than 100 and weigh less than 900 boys of the same age.

3 How many of 1000 normal children are represented in the area above the 75th percentile?

4 How many of 1000 normal children are represented below the 90th percentile?

5 What is the difference in weight between an 18-month-old baby girl on the 5th percentile and an 18-month-old baby girl on the 95th percentile?

6 Compare the overall weight gain of a girl whose growth rate remains on the 5th percentile curve for the first three years of her life and that of another girl who remains on the 75th percentile curve for her first three years.

Being in a higher or a lower percentile does not necessarily mean that a child is healthier or has a growth or weight problem. A baby who is in the 5th percentile can be just as healthy as a baby who is in the 95th percentile. Ideally, each child's growth rate and weight gain will be in proportion. A healthy child usually stays on a certain percentile growth line.

7 Use the boy's and girl's growth charts to decide which of the following children may have health problems. Explain your answers.

a A girl who stays on the 10th percentile line until 6 months old and then moves to the 95th percentile.

b A girl who weighed 5 kg at birth and 7 kg at 12 months.

c A girl who weighed 5.4 kg at 3 months and 7.2 kg at 6 months.

d A girl who weighed 10.2 kg at 12 months and 15 kg at 2 years of age.

e A boy who stays on the 50th percentile line until 18 months and then drops to the 10th percentile.

f A boy whose birth weight was 3 kg, at 3 months was 7 kg and at 6 months 10 kg.

g A boy who weighed 3500 g at birth and 10.2 kg at 12 months of age.

h A boy who weighed 9.5 kg at 12 months and 15 kg at age 2 years.

8 Growth charts exist for children over 3 years of age and for adults. Based on your estimate of the average 'weight-for-age' of someone your age, do you think you would be graphed on the 10th, 25th, 75th or 90th percentile line?

9 How many of the 1000 normal children in your age sample would that mean were heavier than you?

10 Do you consider that your weight and height are in proportion? Explain.

Lesson 4 Growing heavier

It is common for babies to drop between 5% and 10% of their birth weight soon after the birth. Once settled into a feeding pattern an infant is expected to double their birth weight by five months, be three times their birth weight by age one and almost four times their birth weight by 24 months.

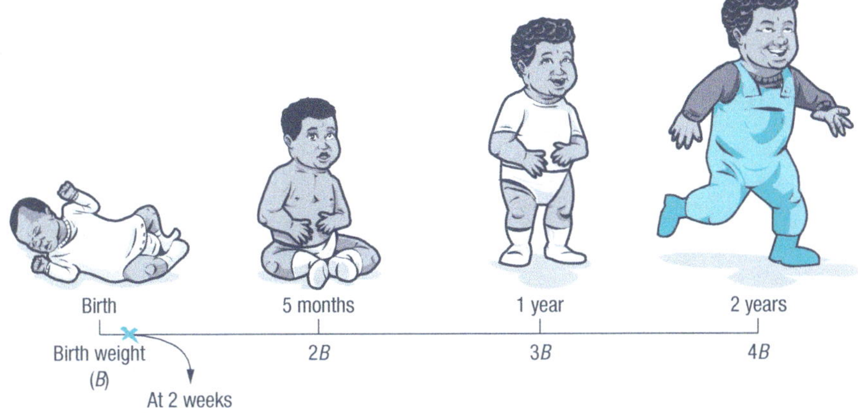

1 Assuming their growth rate was normal, calculate the expected weight of:

a a 1-year-old infant whose birth weight was 2600 g
b a 5-month-old infant who weighed 2.75 kg at birth
c a 1-year-old child whose birth weight was 3200 g
d a 2-year-old child whose birth weight was 2.8 kg
e a 5-month-old who was born weighing 3100 g
f a 2-year-old who weighed 2450 g when born

2 Calculate the birth weight of these 1-year-old children if they had grown at a steady pace yet only increased their birth weight by 50% instead of doubling it as expected. Their current weights are:

a 2 kg b $2\frac{1}{2}$ kg c $2\frac{1}{4}$ kg
d $1\frac{3}{4}$ kg e $3\frac{1}{8}$ kg f $2\frac{2}{3}$ kg

Challenge

Calculate the weight at birth of the 1-year-old children in question 2 if they had grown at a steady pace yet increased their birth weight by 75% instead of doubling it as expected.

A newborn weighing 2400 g might lose 120–240 g (between 5% and 10% of their birth weight) before settling into a normal growth pattern. Their weight would be expected to be within the range of 2160–2280 g during that time.

3 Calculate the weight range of the following newborns if they lost 5–10% of their birth weight.

a birth weight 2500 g b birth weight 1900 g
c birth weight 2250 g d birth weight 1.95 kg
e birth weight 3.1 kg f birth weight 2.35 kg

Earlier this year you learned about ratios. The growth rates described at the start of this lesson can be viewed as ratios.

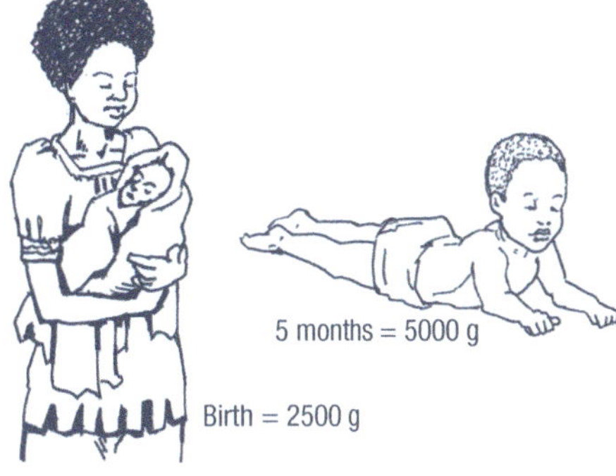

By 5 months an infant usually doubles their birth weight; this is a ratio of 2:1.

4 Convert the following statements to a ratio:

a by age 1 year increased their birth weight three times

b by age 2 years increased their birth weight four times

c by age 1 week decreased their birth weight by 10%

d by age 2 weeks decreased their birth weight by 5%

e by age 1 year increased their length by 50%

f by age 5 months increased their length by 30%

5 A woman is told to mix her baby's milk using a ratio of 2:3 (being two parts cow's milk to three parts water). How much water should she add to a bottle containing the following quantities of milk?

a 20 mL

b 200 mL

c 150 mL

d 0.5 L

e $\frac{1}{4}$ L

f 125 mL

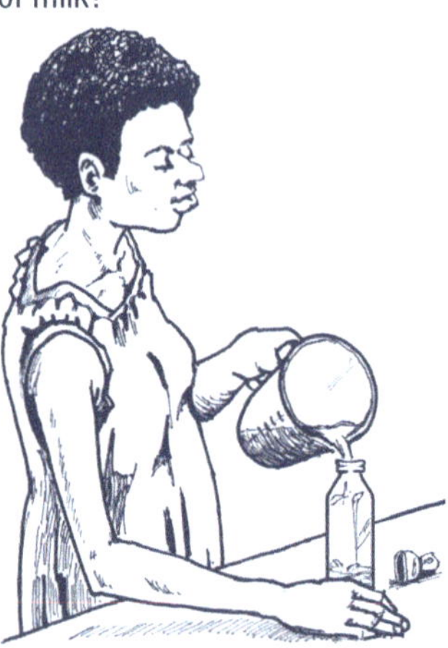

6 A baby's fruit juice is mixed using the ratio 1:2:5 (being 1 spoonful of tonic, 2 spoonfuls of juice concentrate and 5 spoonfuls of water).

a How many spoonfuls of concentrate are needed for 10 spoonfuls of water?

b How many spoonfuls of tonic are needed for 8 spoonfuls of concentrate?

c How many spoonfuls of water are needed for 6 spoonfuls of tonic?

d How many spoonfuls of concentrate are needed for 25 spoonfuls of water?

e How many spoonfuls of tonic are needed for 10 spoonfuls of concentrate?

f How many spoonfuls of water are needed for 12 spoonfuls of tonic?

g What amount of each ingredient is contained in 64 spoonfuls of mixture?

Growth hormone levels naturally rise during puberty in humans. This results in adolescents often experiencing a 'growth spurt' or rapid increase in both their height and weight. The growth spurt begins at about age 11 in boys and age 9 in girls but varies widely between individuals. The average growth spurt lasts 24–36 months.

7 Calculate the weight of an adolescent who:

a weighed 42.5 kg at age 12 and gained 9.25 kg in the next 2 years

b began puberty weighing 38 kg and increased her weight by 25%

c began puberty weighing 41 kg and increased his weight by one-third

d weighed 49 kg at age 13 and increased her weight 15% in the first year of puberty and then gained a further 4.25 kg in the second year

e was 10% heavier each year for 3 years of puberty having started with a weight of 33 kg

f weighed 35 kg at age 11 and by the end of puberty had increased that weight by a ratio of 2:3

g increased their weight during puberty by a ratio of 4:5 from a starting weight of 40 kg

8 a Write an estimate of your birth weight and use that, together with your current weight, to write a sentence describing how many times heavier you are now than your estimated weight at birth.

b Write a ratio to describe the relationship between your estimated birth weight and your current weight.

Challenge

Calculate the percentage increase in your weight from birth until now.

Lesson 5 Growing taller

Babies grow rapidly from birth to 24 months after which their growth rate slows. Infants usually increase their length by about 30% by the first 5 months, and at one year an infant's height has increased by 50%.

1 If a newborn measures 50 cm, what length would you expect her to be at:

a five months

b one year?

2 If a 12-month-old boy is 80 cm tall and has shown a normal growth rate since birth, what length would you expect he was at:

a five months

b at birth?

3 Copy and complete the table to show the expected length of each child at both ages.

Birth length	43 cm	44 cm	46 cm	49 cm	52 cm	55 cm
Expected length at 5 months						
Expected length at 12 months						

4 Use the data you generated in question 3 to create a graph that a parent could use to work out what length they could expect their child to be at 5 months and at 12 months. The graph should have separate lines for the two sets of data.

It is also possible to use fractions when making calculations to predict the expected length of a child at 12 months based on their birth weight.

Help Box

Some fraction calculations can be done more easily than others.

For example, $10\frac{1}{2} \times \frac{1}{2} =$

The commutative law of multiplication means this equation is the same as $\frac{1}{2} \times 10\frac{1}{2} =$

This equation can also be read as $\frac{1}{2}$ of $10\frac{1}{2} =$

The answer can be calculated mentally: $\frac{1}{2}$ of $10\frac{1}{2} = 5\frac{1}{4}$

5 Solve these fraction problems mentally.

a $8 \times \frac{1}{4}$

b $12\frac{1}{2} \times \frac{1}{2}$

c $6\frac{1}{4} \times \frac{1}{2}$

d $14\frac{2}{3} \times \frac{1}{2}$

e $20\frac{1}{3} \times \frac{1}{2}$

f $9\frac{3}{4} \times \frac{1}{3}$

g $15\frac{3}{4} \times \frac{1}{3}$

h $13\frac{1}{2} \times \frac{1}{2}$

i $11\frac{1}{4} \times \frac{1}{2}$

6 Mentally calculate the weight these children will be when aged 1 year if they increase their birth weight by 50% ($\frac{1}{2}$).

a birth weight 3 kg

b birth weight $2\frac{1}{2}$ kg

c birth weight $2\frac{1}{4}$ kg

d birth weight $1\frac{3}{4}$ kg

e birth weight $3\frac{1}{8}$ kg

f birth weight $2\frac{2}{3}$ kg

Help Box

Sometimes fractions can be divided mentally by substituting the term 'how many' for the division symbol.

For example, $2\frac{1}{2} \div \frac{1}{4} =$

This equation can also be read as $2\frac{1}{2}$ *how many quarters*?

The answer can be calculated mentally: $2\frac{1}{2} \div \frac{1}{4} = 10$

7 Solve these fraction problems mentally.

a $5 \div \frac{1}{4}$ b $8\frac{1}{2} \div \frac{1}{2}$ c $5\frac{2}{3} \div \frac{1}{3}$ d $10\frac{1}{2} \div \frac{1}{4}$ e $6\frac{1}{3} \div \frac{1}{3}$

f $3\frac{3}{4} \div \frac{1}{4}$ g $6\frac{1}{10} \div \frac{1}{10}$ h $2\frac{1}{2} \div \frac{1}{8}$ i $3\frac{1}{5} \div \frac{1}{10}$

8 Solve these problems mentally by first converting the decimal fraction to a vulgar fraction.

a $4.5 \div \frac{1}{2}$ b $6\frac{1}{4} \div 0.25$ c $5\frac{1}{2} \div 0.5$ d $10.75 \div \frac{1}{4}$ e $8\frac{1}{10} \div 0.1$

f $11.5 \div \frac{1}{2}$ g $6.2 \div \frac{1}{5}$ h $2\frac{1}{8} \div 0.125$ i $1\frac{2}{5} \div 0.1$

In the previous lesson you learned that growth hormones cause a growth spurt at puberty. Growth in height at this time accounts for about 20% of an adolescent's final height as an adult, and averages 23–28 cm in females and 26–28 cm in males.

9 Write down your current height and calculate:

a your height if you increased in height by 20%. (Of course you may already be into your growth spurt, so this figure is unreliable.)

b your height range if your growth spurt was equal to that stated above for your gender. (Of course you may already be into your growth spurt, so this figure is unreliable too.)

Help Box

Predicting the adult height of a child is difficult, however some health organisations use the 'mid-parent' formulae to make rough estimates.

For girls: $\frac{(\text{father's height} - 13\text{ cm}) + \text{mother's height}}{2}$

For boys: $\frac{(\text{father's height} + 13\text{ cm}) + \text{mother's height}}{2}$

10 Copy the tables into your workbook and use the correct formula to predict each child's adult height.

a

Girl's name	Betty	Mary	Jenny	Alice
Father's height	173 cm	1.7 m	167 cm	1.83 m
Mother's height	156 cm	1.6 m	163 cm	168 cm
Girl's predicted adult height				

b

Boy's name	Otto	Mino	Joseph	Rodney
Father's height	173 cm	1.7 m	167 cm	1.83 m
Mother's height	156 cm	1.6 m	163 cm	168 cm
Boy's predicted adult height				

Lesson 6 Body Mass Index

Body Mass Index (BMI) is an approximate measure of a person's body fat based on their weight and height. Being underweight or overweight can cause health problems. Generally differences between people of the same age and sex are due to body fat, however some exceptions to this mean that the BMI should only be used as a guide.

Help Box

To calculate your BMI you need to know:

Your weight (mass) in kilograms

Your height in metres

BMI is calculated by dividing your weight (in kg) by your height (in m^2).

For example: $\frac{\text{Weight (in kg)}}{\text{Height (in m}^2\text{)}}$ $\frac{70}{1.7^2} = \frac{70}{2.89}$

So the BMI $= 24.22$

1 Use the formula to calculate the BMI of the following adults.

- a height 1.8 m, weight 86 kg
- b height 1.7 m, weight 90 kg
- c height 173 cm, weight 74 kg
- d height 159 cm, weight 64.5 kg
- e height 160.5 cm, weight 72.3 kg

The table shows what an adult's BMI suggests about their body size.

BMI < 19.9	Underweight
BMI 20.0–24.9	Healthy weight range
BMI 25.0–29.9	Overweight (pre-obese)
BMI 30.0–34.9	Moderately obese
BMI 35.0–39.9	Severely obese
BMI > 40.0	Very severely obese

2 Use the table to answer the questions.

- a What description applies to an adult with a BMI of 27.4?
- b List three BMI ratings within the healthy weight range.

3 List the BMIs you calculated for question 1 and, beside each, write the category that person fits into.

The healthy weight range for adults of 20.0 to 24.9 is not a suitable measure for children up to the age of about 20 years, because as children grow the amount of body fat changes and so does their BMI. BMI calculations for children and adolescents should be compared against age and gender graphs.

This graph shows the average BMI for girls between the ages of 11 and 20 years. Any BMI between the 5th percentile and the 85th percentile is considered within the normal range for a female child.

4 Use the graph to answer the questions.

a What is the minimum BMI within the healthy range for an average 14-year-old girl?

b What is the maximum BMI within the healthy range for an average 14-year-old girl?

c If an 11-year-old girl had a BMI of 22, would she be described as underweight, normal, overweight or obese for her age?

d If a 16-year-old girl had a BMI of 22, how would she be described?

e What range of BMI is considered 'normal' for the average 17-year-old girl?

f What is the minimum BMI at which an average 13-year-old girl would be considered obese?

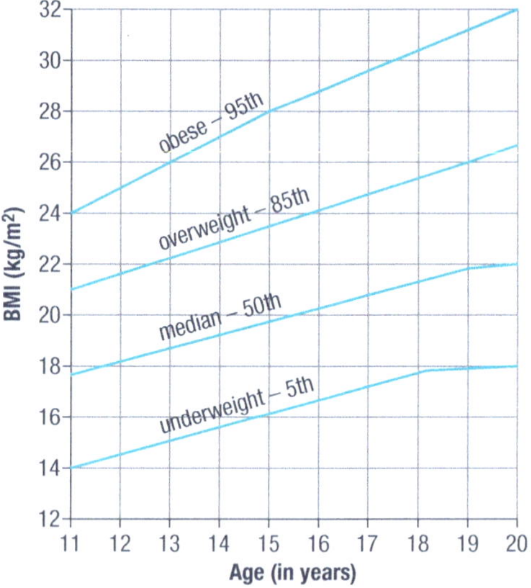

This graph shows the average BMI for boys between the ages of 11 and 20 years. Any BMI between the 5th percentile and the 85th percentile is considered within the normal range for a male child.

5 Use the graph to answer the questions.

a What is the minimum BMI within the healthy range for an average 12-year-old boy?

b What is the maximum BMI within the healthy range for an average 12-year-old boy?

c If a 13-year-old boy had a BMI of 20, would he be described as underweight, normal or obese for his age?

d If a 15-year-old boy had a BMI of 20, how would he be described?

e What range of BMI is considered 'normal' for the average 19-year-old boy?

f What is the minimum BMI at which an average 16-year-old boy would be considered obese?

6 Calculate the BMI for these children and use the above graphs to decide if they are underweight, normal, overweight or obese compared to the average child of that age.

a a 14-year-old female, 1.3 m tall who weighs 40 kg

b a 15-year-old male, 1.5 m tall who weighs 42 kg

c an 11-year-old female who is 145 cm tall and weighs 28 kg

d a 19-year-old male, 185 cm tall and weighing 70 kg

7 a Measure your own height and weight and calculate your BMI.

b Compare your BMI to the relevant graph. Write a sentence commenting on your actual weight and what the graph shows as the normal range for someone your age.

c What does your own BMI suggest about your current eating and exercise habits?

Lesson 7 Pulse rate and percentage error

A person's heart rate (often called pulse rate) is the number of times their heart beats each minute. When resting an adult's heart rate varies between 55 and 80 beats per minute, although a range of factors including the person's age, gender and fitness influence this.

You can feel your heart beat if you hold your fingers against the main artery in the side of your throat.

1 Work with another student of about your age to calculate your pulse rate. One student times a 20-second block while the other counts how many times their heart beats in that time. Multiply the number of heart beats for 20 seconds by three to find your pulse rate per minute.

2 Copy the table into your workbook and complete it by collecting and recording both your own and your partner's pulse rate, as well as the pulse rate of four other students who are your age.

Name						
Resting pulse rate						

3 What is the range of the six pulse rates?

4 What is the mean pulse rate of the six students?

In some situations it is important to decide what degree of accuracy applies to calculations. For example, a group of 12-year-old males counted the number of breaths they took in a minute and recorded their results in a table. From their results they decided they could estimate that an average 12-year-old male takes 15(± 2) breaths per minute. (The ± symbol means plus or minus.)

Name	Breaths p/min
Rodney	14
Sam	17
Wesley	16
Andrew	13
Total	**60**
Mean	**15**
Range	**4**

5 Look back at the information about pulse rates in questions 2, 3 and 4. Copy and complete the following sentence into your workbook using the information you collected.

The average pulse rate of the six students is ______ (± ____).

Challenge

When exercising, your pulse rate increases. The ideal training heart rate per minute is found using the formula:

$$(220 - \text{your age}) \times 0.75$$

Calculate the average ideal training rate of the six students including a ± error margin.

6 Copy and complete the following. The first one has been done as an example for you.

a 12 (± 3) can be any answer within the range 12 – 3 and 12 + 3 = (9 to 15)

b 15 (± 2) can be

c 32 (± 6) can be

d 3.7 (± 0.4) can be

e 11.2 (± 0.3) can be

f 1.1 (± 0.05) can be

g 2.04 (± 0.05) can be

Help Box

The margin for error can also be expressed as a percentage.

For example, 230 ± 10% means any number within 10% of either side of 230 (the range 207 to 253).

7 Copy and complete the following. The first one has been done as an example for you.

a 120 (± 10%) can be any answer within the range 120 – 12 and 120 + 12 = (108 to 132)

b 120 (± 5%) can be

c 140 (± 10%) can be

d 160 (± 5%) can be

e 45 (± 20%) can be

f 250 (± 25%) can be

g 80 (± 15%) can be

h 250 (± 6%) can be

Help Box

To calculate the percentage error, use the formula:

$$\% \text{ error} = \frac{(\text{estimate} - \text{actual value})}{\text{actual value}} \times 100$$

For example, an average healthy 20-year-old male estimates his pulse rate will be 150 after a 20-minute run. His actual pulse rate is 140.

Subtract the actual value from the estimated rate: 150 – 140 = 10

Divide the result by the actual value: $\frac{10}{140} = 0.07$

Express as a percentage: 0.07 × 100 = 7%

The final percentage error in the estimation is 7%.

8 Calculate the percentage error in the following estimations.

a estimated value 100, actual value 80

b estimated value 50, actual value 48

c estimated value 150, actual value 120

d estimated value 90, actual value 60

e estimated value 75, actual value 60

f estimated value 30, actual value 24

g estimated value 16, actual value 14

Lesson 8 Estimating measurement

Many health experts recommend that to remain healthy the minimum number of steps that a normal adult should take in one day is 10 000.

1 a Estimate the distance you would walk if you took 10 000 steps. Include a ± amount to show the degree of accuracy you expect.

b Explain your thinking to make the estimation.

2 Estimate the distance you step in one normal walking stride. Include a ± amount to show the degree of accuracy you expect.

3 Measure your normal walking stride.

- Make a mark on the ground.
- Place your big toe against the mark and then step off with the other foot and take 10 steps at your normal stride.
- Mark where your toe finishes on the tenth step.
- Measure the distance between the two marks and divide the answer by 10 to calculate one normal step length.

4 Write a sentence describing how the actual length of your step related to your estimation.

5 a Based on the distance you cover in one normal walking step, calculate the total distance that you would walk if you took 10 000 steps a day.

b What distance do you estimate that you walk on a normal school day at the moment? Explain your reasoning.

Challenge

Calculate how much further you would walk on a normal school day if you increased your normal stride by 5 cm but still took 10 000 steps. How much further would this be over a 40-week school year?

6 a Estimate the length of the classroom and include a ± amount to show the degree of accuracy you expect.

b Explain how you worked out your estimation.

7 Measure the actual length of the classroom and write a comment about how well you estimated.

It is usually much harder to estimate the length of a curved line than a straight line. As with anything else, more practice can improve estimation skills.

8 a Collect some round bottles and tins and estimate their circumferences including a ± amount to show the degree of accuracy you expect.

b Explain your reasoning when trying to estimate measurements of circumference.

c Use string and a ruler to find the actual measurements.

d Write a comment about how well you estimated.

9 a Estimate how long it would take you to walk heel to toe, heel to toe across the classroom.

b Ask a friend to time you and write down the actual time taken.

c Calculate the percentage error of your estimation.

10 a Put some water into an unmarked container and estimate the amount.

b Describe how you worked out your estimate.

c Pour the water into a measuring jug and write down the actual amount.

d Calculate the percentage error of your estimation.

11 Work with a classmate to compare your estimation skills. Below is a list of items to be measured. Each student should write an estimate of the answers first, then both work together to measure the items. Write the actual measurements in your workbook and calculate your own percentage error for each. When all the items have been measured, each student should add up their percentage errors. The student with the smallest total wins.

a the height of a door

b the circumference of a circular object

c the length of a desk or table

d the capacity of a cup (ask the teacher for one)

e the mass of your workbook

f the width of a window

Lesson 9 How hot is it?

Temperature is the measure of how hot things are. In Papua New Guinea temperature is measured in degrees Celsius. Water freezes at 0°C and boils at 100°C.

A normal, healthy, resting adult has a body temperature of 37°C. The body is very good at keeping its temperature within the narrow range from 36.1°C to 37.8°C in spite of large variations in temperature outside the body. An abnormally low body temperature (hypothermia) or an abnormally high body temperature (fever) can be serious, even life-threatening.

1 Read the clinical thermometers and record the temperatures to the nearest tenth of a degree.

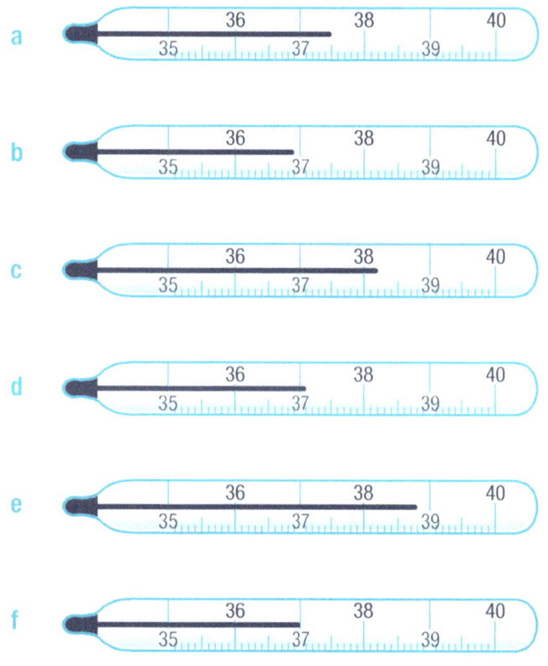

2 List the temperatures in question 1 that might indicate a person is unwell. Explain your choices.

3 Draw clinical thermometers that show the following temperatures.

a 35.9°C b 37.4°C c 36.3°C
d 38.2°C e 39.1°C f 36.8°C

4 An athlete's body temperature was taken every 5 minutes during a 30-minute workout. The athlete started at 36.3°C and increased his body temperature by 0.3°C in the first 5 min, 0.5°C for each of the next two readings, then increased by 0.2°C for the next reading and a further 0.1°C after that. The athlete's body temperature had dropped 0.6°C by the final reading after he had spent time stretching.

a Draw a graph and plot the athlete's body temperature for the 30 minutes.

b Draw thermometers to show his starting and finishing temperatures.

c What was the difference between his starting and finishing temperatures?

d Was the athlete outside the healthy temperature range at any time? Explain your answer.

Other types of thermometers are used to measure the humidity in the air. Measures of humidity tell us the extent to which air is saturated with moisture at a given temperature. When you get a shock every time you touch something, the air is too dry. However if water is condensing on your windows, the air is too damp (ideal for growing mould and mildew). A simple way to calculate the relative humidity is to use a wet-and-dry thermometer, called a psychrometer and sometimes a wet-and-dry bulb hygrometer.

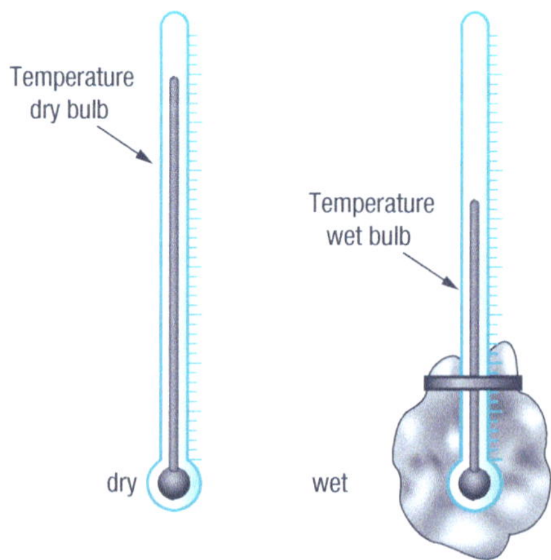

5 Work with another student and follow the steps to find the relative humidity level. Ideally you would use two thermometers, but it is possible to work with one.

Step 1: Measure the air temperature with a thermometer.

Step 2: Wrap cotton wool or cloth around the bulb of a thermometer. Secure it with a rubber band.

Step 3: Dip the wrapped end of the thermometer in water.

Step 4: Wait 10 minutes until the temperature stabilises.

Step 5: Subtract the wet temperature from the dry temperature.

Step 6: Look at the Relative Humidity Table below. The numbers in the left-hand column represent the dry bulb reading. The numbers across the top represent the difference between the wet and dry bulb readings (your answer to step 5).

Step 7: Find where the dry bulb temperature row intersects with the difference column and record the number at that point. The number is the relative humidity of your room as a percentage.

Relative humidity (%)										
Dry bulb (ºC)	**Difference between wet and dry bulb readings**									
	1	**2**	**3**	**4**	**5**	**6**	**7**	**8**	**9**	**10**
10	88	77	66	55	44	34	24	15	6	
11	89	78	67	56	46	36	27	18	9	
12	89	78	68	58	48	39	29	21	12	
13	89	79	69	59	50	41	32	22	15	7
14	90	79	70	60	51	42	34	2	18	10
15	90	80	71	61	53	44	36	27	20	13
16	90	81	71	63	54	46	38	3	2	15
17	90	81	72	64	55	47	40	32	25	18
18	91	82	73	65	57	49	41	34	27	20
19	91	82	74	65	58	50	43	36	29	22
20	91	83	74	67	59	53	46	39	32	26
21	91	83	75	67	60	53	46	39	32	26
22	92	83	76	68	61	54	47	40	34	28
23	92	84	76	69	62	55	48	42	36	30
24	92	84	77	69	62	56	49	43	37	31
25	92	84	77	70	63	57	50	44	39	33

6 Write a few sentences to describe what you notice about the change in the level of humidity as the difference between the wet and dry thermometer increases and as the temperature of the dry thermometer increases.

7 How do you think the humidity of an empty classroom would compare to the humidity at the same time in a classroom full of students? Explain your answer.

Challenge

Repeat the exercise elsewhere in the school predicting what the humidity might be before you take your measurements.

Learning Unit 2 The Past and the Future

Strand: Chance and Data

Statistics	Outcome 8.4.1	Interpret information presented statistically
Sets	Outcome 8.4.2	Use sets to solve problems from real life
Estimation	Outcome 8.4.7	Estimate results of calculations

Strand: Patterns and Algebra

Algebra	Outcome 8.5.2	Recognise and use patterns in processes

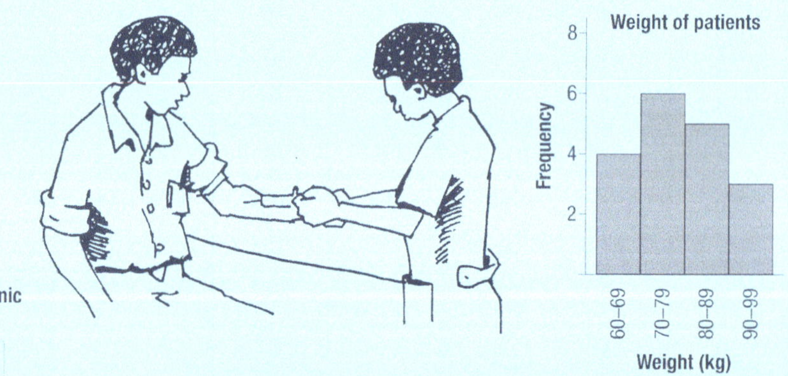

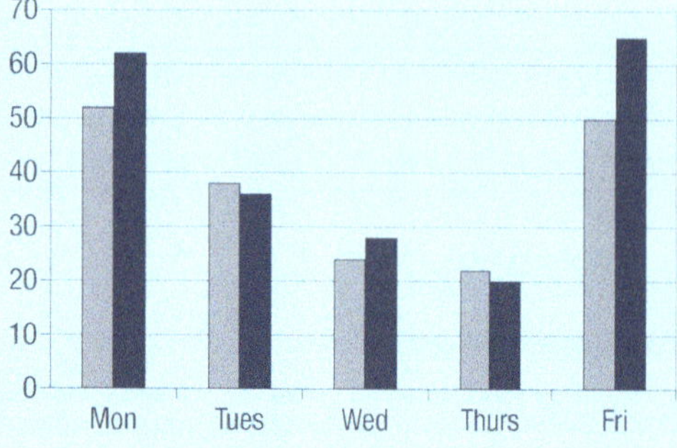

Infant mortality rate		
Year	**1990**	**2006**
Under 5 mortality rate (per 1000)	94	73
Under 1 mortality rate (per 1000)	69	54

Source: UNICEF www.unicef.org

Lesson 1: Introduction	Determining positive and negative influences on health
Lesson 2: How healthy were we then?	Analysing statistical data
Lesson 3: Random samples	Estimating using random sample data
Lesson 4: Making predictions	Making predictions from sample data
Lesson 5: Making sense of it	Interpreting and analysing data Graphing data
Lesson 6: More on statistics	Using frequency tables Calculating relative frequency
Lesson 7: Column graphs and histograms	Graphing discrete and continuous data
Lesson 8: Who belongs where?	Using Venn diagrams to classify sets
Lesson 9: Number patterns at work	Using number patterns and formulae

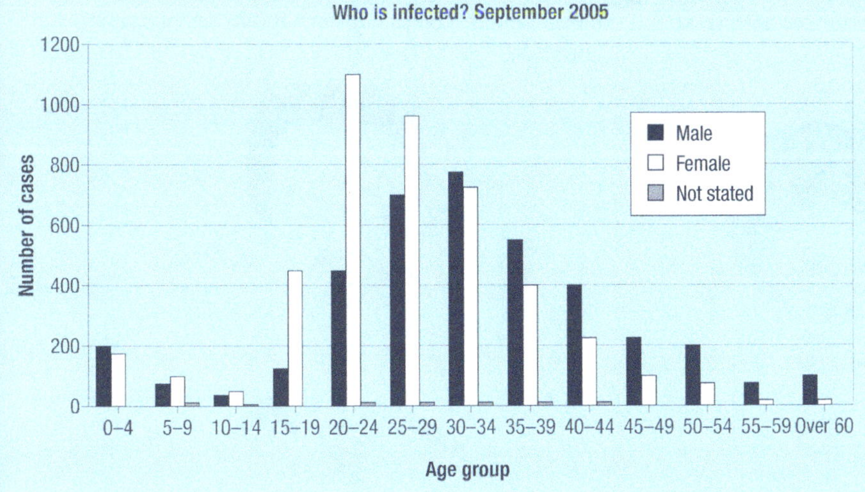

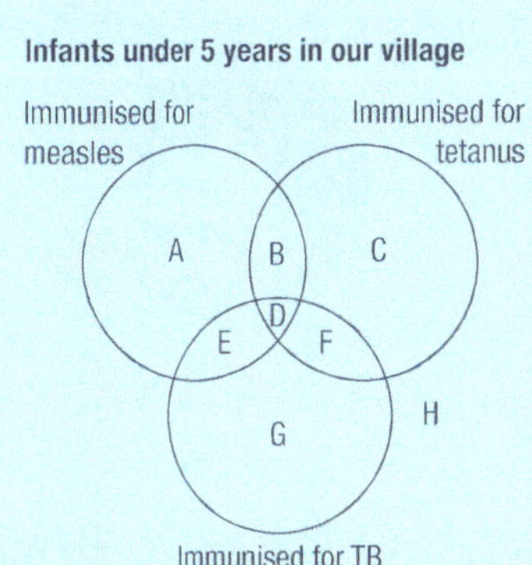

Lesson 1 Introduction

During this unit you will look at some of the data describing the health of the population of Papua New Guinea. Some trends will be identified that highlight high-risk practices as well as the people most at risk from these practices. You will use a random sample of data to make predictions about future health issues.

It is far more effective to look after a nation's health by preventing the spread of infection and disease than it is to have to cure widespread illness. Before people can change their behaviour to reduce their risk of infection and disease, they must understand some basic facts about health, adopt healthy attitudes, practise healthy behaviours and have access to appropriate health services and products.

Each of these illustrations identifies a method of preventing the spread of infection and disease.

1 Describe the preventative methods shown above.

2 List five things in your surroundings that may have a positive effect on your general health, for example, good friends.

3 List five things in your surroundings that may have a negative effect on your general health, for instance, pollution.

4 In a small group, choose one of the negative influences on your general health. Together make a list of actions you can take to reduce and/or prevent this issue from having a negative effect on your growth and development.

Lesson 2 How healthy were we then?

Various international aid organisations collect data about different countries in order to make comparisons and determine where medical assistance will be most needed. These tables show some of the data that has been collected about Papua New Guinea.

Infant mortality rate		
Year	**1990**	**2006**
Under 5 mortality rate (per 1000)	94	73
Under 1 mortality rate (per 1000)	69	54

Source: UNICEF www.unicef.org

Life expectancy		
1970	**1990**	**2006**
43	54	57

Source: www.worldbank.org

Health		
Year	**2000**	**2005**
Population	5.3 mill	5.9 mill
Population growth (annual %)	2.3%	2.0%
Immunisation, measles (% of children ages 12–23 months)	62%	60%

Source: www.worldbank.org

1 a Use a dictionary to find the meaning of 'infant mortality'.

b Read the data presented in the table titled 'Infant mortality rate' and write a few sentences explaining your interpretation of the data.

Data from the Population Reference Bureau (http://www.prb.org/datafinder.aspx) shows the average life expectancy in Papua New Guinea in 2007 was 57 years, with women expected to live to 60 and men to 54.

2 If the data in the table titled 'Life expectancy' are averages only, suggest what the life expectancy for men and for women might have been for each of the three years shown.

3 Predict what the life expectancy for men and for women might be in 2026 if the current trend continues. Explain your reasoning.

4 The table titled 'Health' shows that the population increased from 2000 to 2005 and yet the population growth reduced from 2.3% to 2%. Explain what this means.

5 a What might be the consequences of the data in the 'Health' table relating to measles immunisation?

b Suggest some action that the government might take in response to this data.

The World Health Organization (WHO) presents its data within regional areas containing several neighbouring countries. The following data shows some WHO data for Papua New Guinea and similar data for the broader region in which WHO have classified Papua New Guinea.

Cause of death in infants under 5 years (2000–03)		
	PNG deaths* (%)	**Regional deaths* average (%)**
Total neonatal deaths	100	100
Neonatal causes	35	47
HIV/AIDS	0	0
Diarrhoeal diseases	15	12
Measles	2	1
Malaria	1	0
Pneumonia	18	14
Injuries	2	7
Other	25	18

* Sum does not total 100 due to rounding. Source: www.who.int

6 Use a dictionary to find the meaning of the word 'neonatal'.

7 Use the data in the WHO table to answer the following questions.

a What group of people is described by this data?

b List the three major causes of neonatal deaths in Papua New Guinea.

c Which one of these causes might most easily be reduced with government intervention? Suggest some methods that could be used to achieve the reduction.

d List the three major causes of neonatal deaths for the WHO Region that includes Papua New Guinea.

e Write a few sentences comparing the differences between the data for Papua New Guinea and that for the WHO Region overall.

f Suggest three causes of death that might be categorised as 'other' in the data.

g Draw a column (or bar) graph to represent the two sets of data on one graph.

8 Data from UNICEF (http://www.unicef.org) estimates that 14 000 children under the age of 5 years died in Papua New Guinea in 2006. Based on that estimate and using the data from WHO for question 7, calculate the number of children in Papua New Guinea under 5 years who died from:

a diarrhoeal diseases

b measles

c pneumonia

The first case of HIV infection in Papua New Guinea was detected in 1987. By June 2005, over 12 000 people had been reported to be living with HIV/AIDS. HIV is now the leading cause of adult mortality at Port Moresby General Hospital. (Source: http://www.who.int/hiv/HIVCP_PNG.pdf)

The following data compares the HIV epidemic in Papua New Guinea with the rest of the world.

Prevalence of HIV/AIDS		
	2001 HIV/AIDS population ages 15–24 (%)	2005–06 HIV/AIDS population ages 15–49 (%)
PNG	0.4	1.8
World	–	0.9

Source: Population Reference Bureau: www.prb.org/datafinder.aspx

9 a Write a few sentences to describe what HIV/AIDS means and how it is transmitted.

b How does the data from Papua New Guinea compare to that from the rest of the world?

The graph shows the prevalence of HIV/AIDS in Papua New Guinea by gender and age. Use the data to help you answer the following questions.

10 What does the graph show about the relationship between HIV/AIDS and age in Papua New Guinea?

11 a What does the graph show about the relationship between HIV/AIDS and gender in Papua New Guinea?

b Why do you think that one gender is more likely to test positive for HIV than the other?

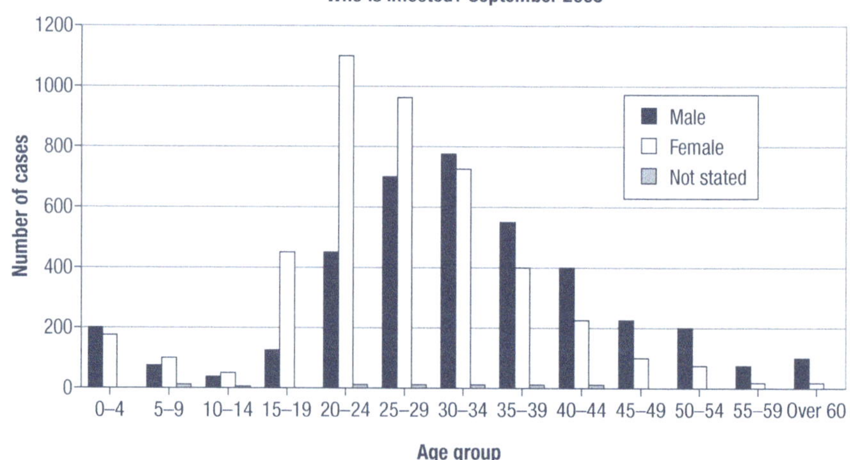

Lesson 3 Random samples

When data is collected from every member of the group being studied, it is possible to make conclusions about the group with some degree of certainty. This graph shows the number of people seen by a doctor at a health clinic during a two-week period. Use the data to help you answer the following questions.

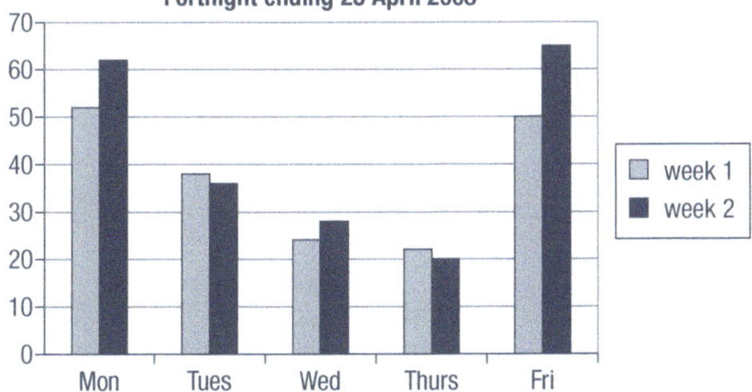

1 List the days of week 1 in order from most busy to least busy.

2 What was the total number of patients seen by the doctor during the two-week period?

3 Calculate the daily mean for week 1 and for week 2.

4 Describe any trends that you notice in the data.

5 Would it be reasonable to make the following statement based on this data: 'Every doctor in Papua New Guinea sees fewer patients on Thursday than any other day of the week?' Explain your answer.

Sometimes the population of interest is too large for us to measure every one of its members, for example, all Grade 8 students in Papua New Guinea, all people attending clinics, all car-owners, and so on. Instead we can choose a random sample from the total population. A **random sample** means every item in the population has an equal chance of being selected. Random samples are not perfect (especially if small) and may not necessarily be a good cross-section of the population.

6 Consider the following example. A flat ruler is thrown and lands numbered-side up. If we used this as our sample, we would conclude that 'when this ruler is thrown, 100% of the time it will land numbered-side up'.

- a When the ruler is tossed 10 times, it lands 7 times with the numbered-side up. What conclusion might we make based on this sized sample?
- b Tossing the ruler 1000 times provides a far more reliable result. The ruler lands 508 times numbered-side up and 492 times with the numbered-side down. What conclusion might we make based on this sized sample?

7 129 out of 2500 students were examined and 15 were found to have skin infections.

- a What was the population size?
- b What was the sample size?
- c What fraction of the sample had a skin infection?
- d Estimate the number of students in the population with a skin infection.

Help Box

To estimate the population with a particular characteristic:

$$\frac{\text{Number with the characteristic in the sample}}{\text{sample size}} \times \text{population size}$$

For example, 15 out of 40 students had an eye test and 3 were found to have an eye problem. Estimate the percentage of the population with an eye problem.

$$\frac{3}{15} \times 40 = \frac{1}{5} \times 40$$

$$= 8 \text{ of the whole population (or } \frac{8}{40} \times 100 \text{ which is 20\% of the population)}$$

So 20% of the population is estimated to have an eye problem.

8 A clinic immunises 300 people against measles each month. Later a random sample of 50 found that two people had contracted measles.

- a What is the size of the population?
- b What size is the sample?
- c Estimate how many people immunised against measles each month contract the disease.
- d Estimate how many people immunised against measles each month do not contract the disease.

9 Collect data from a random sample of students asking the question 'Do you think schooling should be compulsory until age 17 years?'.

- a What is the total school population?
- b What size was your sample?
- c What fraction of the sample agreed with the statement?
- d Estimate the number of students in the school who would agree with the statement based on the random sample results.
- e Comment on the reasonableness of your conclusion based on the size of your sample.

Challenge

A small lake contains only one breed of fish. The rangers caught 200 of the fish on one day, tagged them and released them back into the lake. Two weeks later 100 fish were caught, and of these 12 had tags. Estimate the number of fish in the lake. Explain your thinking.

Lesson  Making predictions

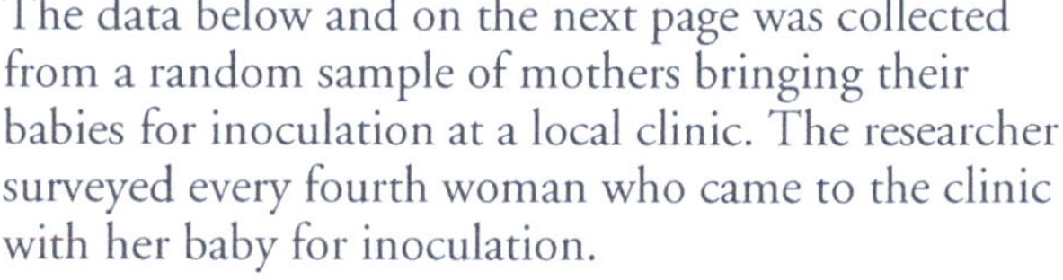

It is important to think about how a sample population is selected if the conclusions are to be generalised to a larger population.

1 How would you randomly select:

- a six students to be on a committee to decide what sporting equipment the school should buy?
- b some groceries to make comparisons between prices at different food stores?
- c some adults to find the amount of time most people spend travelling to work?
- d some tourists to decide what they spend most of their money on in Papua New Guinea?

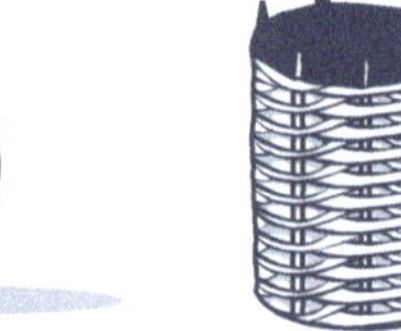

The data below and on the next page was collected from a random sample of mothers bringing their babies for inoculation at a local clinic. The researcher surveyed every fourth woman who came to the clinic with her baby for inoculation.

Age of mother	
15–20	63
21–25	26
26–30	10
> 31	1
Total	100

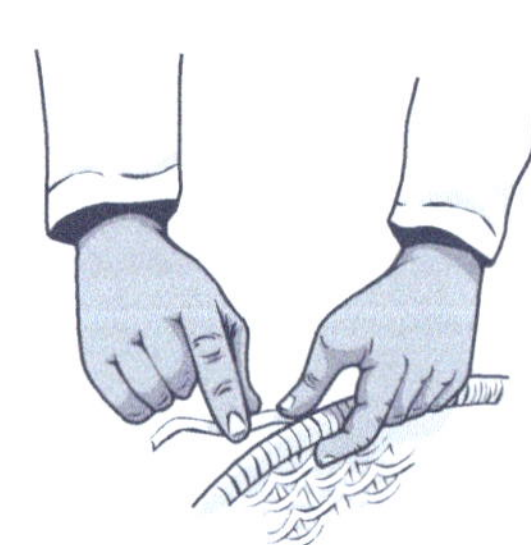

Highest education level received	
Grade 8 or lower	43
Grade 9 or 10	45
Grade 11 or 12	9
Tertiary studies	3
Total	100

2 What fraction of the total number of women who came to the clinic to have their babies inoculated was selected for the sample survey?

3 What other information might you need to know about this random sample before you could generalise the results to describe the total population of mothers of babies in Papua New Guinea?

4 Based on the random sample, which of the following conclusions would seem reasonable?

a At this clinic, most mothers of babies are under 20 years of age.

b The number of mothers in Papua New Guinea with a Grade 8 education is about the same as the number with a Grade 10 education.

c In Papua New Guinea only 3% of mothers have undertaken tertiary studies.

d 10% of the women in this survey were over 26 years of age.

5 For each statement in question 4 that you decided was not reasonable, explain your reasoning.

This graph shows data collected from a random sample involving more than half of the people attending a village health clinic during one week.

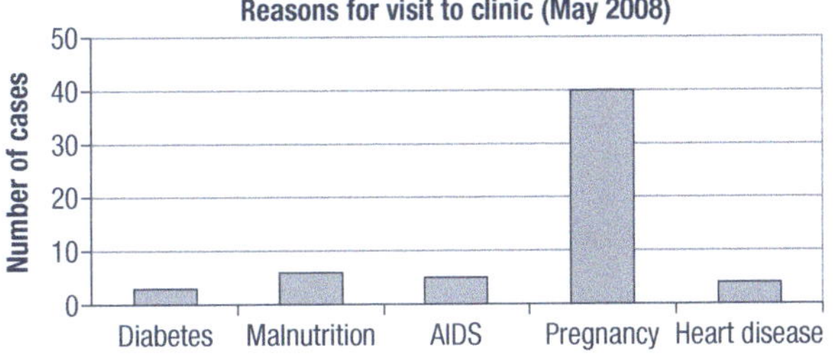

6 Based on the data in the graph, what public facilities might the local council need to spend more money on in the future? (You could consider roads, hospitals, inoculation programs, cemeteries, schools, and so on.) Explain your thinking.

7 Predict some possible changes to the village population in 20 years from when this data was collected. Explain reasons for this change based on the data in the graph.

8 a Collect data from a random sample of students about where they hope to live in 20 years' time. Your random sample could be, for example, every second person who walks past you at lunchtime.

b Display your data, including a description of the method you used to make sure the sample was random.

c Make some predictions about the future spread of the population in Papua New Guinea based on your random sample.

d Describe some characteristics of the random sample that may mean the predictions might not be reasonable.

Challenge

Make some predictions about the future using a random sample of students and your own research question.

9 Write a paragraph to discuss how reasonable the following suggestion is. 'To find out the percentage of people who believe that Port Moresby should build a larger airport it would be a good idea to ask people shopping at the village market.'

10 Suggest the group of people it might be most appropriate to survey if the school was considering holding a monthly craft market to raise funds. Explain your answer.

Lesson 5 Making sense of it

Data can be presented in many ways. The type of data collected helps us to decide how best to represent it.

1 Describe an example of information that would be suitable to present in a:

a column graph or bar graph b line graph c pictogram d pie chart

The bar graph below shows the increase in reported cases of HIV/AIDS in Papua New Guinea since 1987. Use the information in the graph to help you answer the following questions.

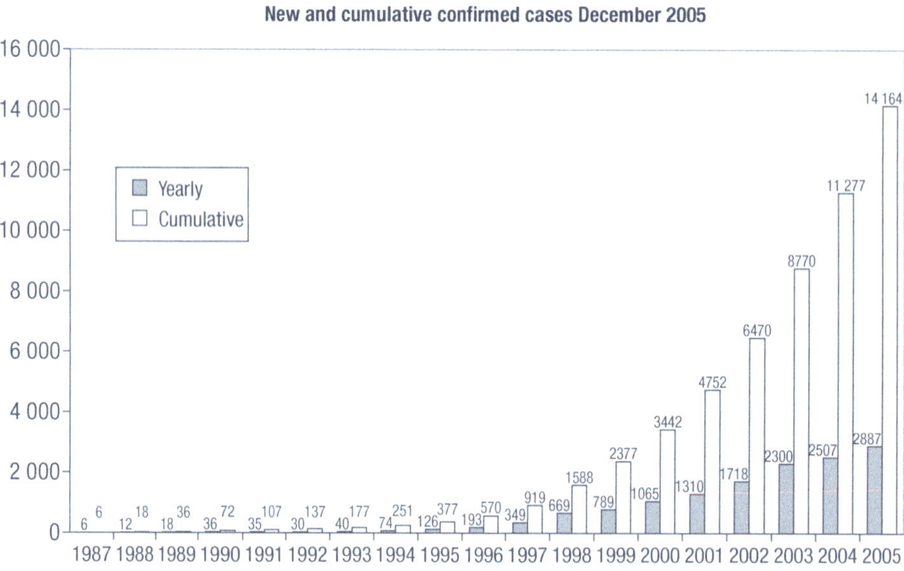

2 a Why are the bars presented in pairs on the graph?

b What number of HIV/AIDS cases was reported in 2001?

c What was the cumulative total of reported cases in 2003?

d How many more cases were reported in 2005 than in 1987?

Help Box

Earlier this year you learned to calculate percentage increase.

$$\text{Percentage increase} = \frac{\text{increase}}{\text{original number}} \times 100$$

3 Use the data in the HIV/AIDS graph to calculate:

a the percentage increase in reported cases from 1987 to 2005

b the percentage increase in the cumulative total of reported cases from 1987 to 2005

4 If the trend in reported cases of HIV/AIDS has not altered since this graph was created, predict:

a the number of cases that were reported in 2008

b the cumulative total of reported cases in 2008

The Feelgood Clinic runs three different vaccination programs. Mothers are encouraged to have their infants immunised against three different diseases. This graph shows the cases reported at the clinic of diseases preventable by immunisation.

Feelgood Clinic—Diseases preventable by immunisation

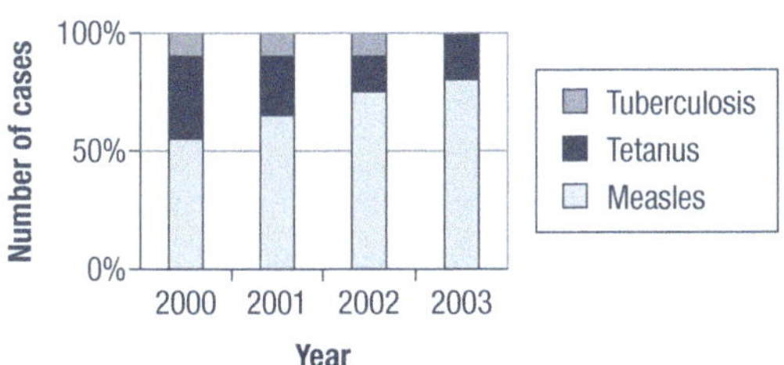

5 a According to the graph, what are three preventable diseases?

b Which of the three diseases does the immunisation program seem to be affecting the most? Explain your answer.

c What conclusions can you make about the measles vaccination program being run by this clinic?

6 Write a few sentences to explain the benefits of immunising infants against preventable diseases both to the infant and to the broader community.

Drinking alcohol over a long period of time can have serious effects on the health of the drinker and their family and friends. This table shows the level of risk involved in relation to the number of drinks per week.

	Low risk	Dangerous	Harmful
Male	Fewer than 28	28–42	More than 42
Female	Fewer than 14	14–28	More than 28

7 Represent the data in a line graph showing the level of risk for women on one line and men on another line.

8 Sketch a pie chart to roughly display the following data relating to government expenditure.

Government expenditure 1995	
Sector	**Expenditure (millions of kina)**
General public service	220
Defence	60
Education	220
Health	90
Social security, welfare	10
Housing	10
Economic services	220
Industry	80
Electricity	30
Transport	50
Other	10
Total	1000

9 Why do you think such a large portion of government expenditure goes towards education?

10 List three jobs that would be paid from the portion described as 'general public service'.

Challenge

Redraw the table of government expenditure to reflect how you think the money should be shared among the items listed. Write a few sentences to explain the major differences between the two tables and why you feel the money should be spent differently.

Lesson 6 More on statistics

Organising data into tables and graphs makes it easier to make comparisons between items and predictions about the future. Statistical data is often presented in a table that shows individual values as well as the number of times each value occurs. This form of table is called a **frequency table**.

Help Box

The values (7, 9, 8, 5, 8, 7, 5, 6, 7, 7, 5, 7) are represented in this frequency table.

Value	Tally	Frequency
5	III	3
6	I	1
7	𝍸	5
8	II	2
9	I	1

1 A survey was conducted on the age of women having their first baby at a particular hospital during a 3-month period.

17 15 25 19 17 20 16 17 20 24
21 20 19 21 19 22 17 22 21 20
19 21 22 24 19 18 19 22 23 16
21 24 15 17 20 21 19 18 20

a Create a frequency table of the data.

b What was the most common age of the women in the study?

c What percentage of women in the study were under 20 years of age?

Help Box

Earlier you have learned about different ways of analysing data.

The **mean** is calculated by dividing the sum of the values by the number of items in the set.

The **mode** is the value that occurs most frequently in a set of data.

The **median** is the middle value in an ordered set of data.

2 Using the data in question 1, calculate:

a the mean b the mode

c the median of the data.

3 The following list shows the weight (in kg) of each student in a particular class.

71 85 64 48 53 64 72 71 49 60
56 72 49 56 61 72 56 61 60 58

a Create a frequency table to represent the data more appropriately.

b How many students weigh more than 60 kg?

c What percentage of students weigh more than 60 kg?

d What is the students' mean weight?

When the data involves a large range of values, it is often more appropriate to group the data into classes when creating a frequency table.

4 Some students' general knowledge test scores were entered into the following frequency table.

Test score	Tally	Frequency
0–19	III	3
20–39	𝍸	5
40–59	𝍸 𝍸 I	11
60–79	III	3
80–99	I	1

a How many students scored less than 40 on the test?

b What percentage of students scored 60 or more on the test?

c Within what test score group were the highest number of results?

Frequency tables often include a column showing the relative frequency of a certain value. The relative frequency of a certain value is written as a decimal fraction. It is an approximation of the probability of an event occurring.

Help Box

$$\text{Relative frequency} = \frac{\text{frequency}}{\text{sample size}}$$

For example, the number of people attending a medical clinic each day during one fortnight was:

7, 5, 6, 8, 5, 10, 5, 7, 9, 7, 8, 8, 5, 10

Score	Tally	Frequency	Relative frequency
5	IIII	4	4 ÷ 14 = 0.29
6	I	1	1 ÷ 14 = 0.07
7	III	3	3 ÷ 14 = 0.21
8	III	3	3 ÷ 14 = 0.21
9	I	1	1 ÷ 14 = 0.07
10	II	2	2 ÷ 14 = 0.14
	Totals	**14**	**0.99**

The relative frequency of 0.29 in the table means that there is a 29% probability that the number of people attending the clinic on any single day will be 5.

5 Copy the following frequency table into your workbook and complete it by calculating the relative frequency of each value.

Score	Frequency	Relative frequency
15	6	
16	9	
17	16	
18	11	
19	6	
20	2	
Totals		

6 Some footballers competed in a kicking competition. The distances they kicked are recorded in the table.

Distance (m)	25–29	30–34	35–39	40–44	45–49	50–54	55–60
No. of footballers	4	9	16	28	27	8	4

a Copy the table into your workbook and make a column for relative frequency.

b How many footballers kicked less than 40 m?

c What percentage of footballers kicked at least 45 m?

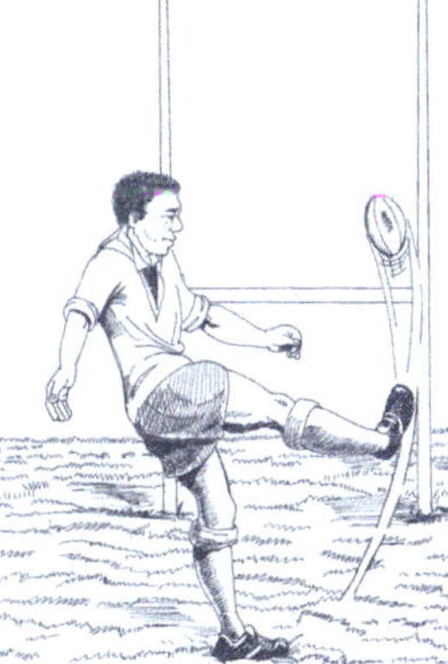

7 The fitness level of a group of students was scored on a scale from 1 to 10. The scores were:

9, 7, 6, 8, 5, 9, 8, 7, 8, 7, 5, 7,
6, 6, 5, 9, 8, 7, 8, 6, 7, 7, 8

a Draw a frequency table and include a column of relative frequency.

b What percentage of students scored 8 or above?

c What percentage of students scored less than 7?

Lesson 7 Column graphs and histograms

Column graphs and histograms are two similar ways of displaying data. **Column graphs** have gaps between the columns to show that the data is discrete (individually different). Column graphs are sometimes called bar graphs, and can be vertical or horizontal. For example, data about the number of babies inoculated against different diseases could be shown on a bar graph because the diseases have no numerical value and their order on the graph does not matter.

Histograms have no gaps between the columns and are used when data is continuous (connected without a break). For example, data about how many students in the grade are 120–139 cm, 140–149 cm and 150–159 cm in height would be shown on a histogram. A histogram shows the frequencies of different numerical values within a data set.

Help Box

The number of babies inoculated against different diseases would be shown on a column graph because the data is discrete.

Column graph

discrete data

The height of a group of students would be shown on a histogram because the data is continuous.

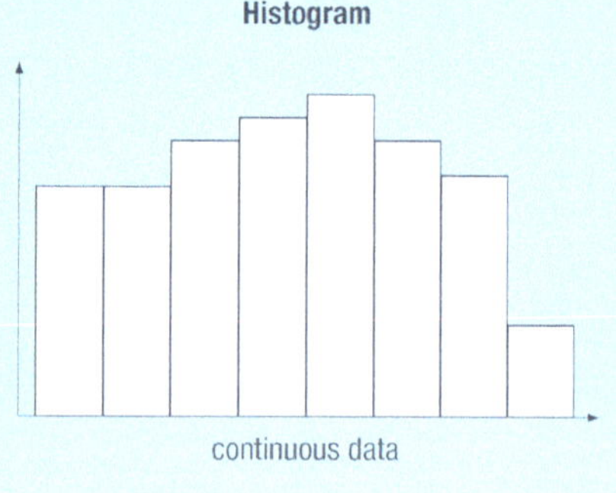

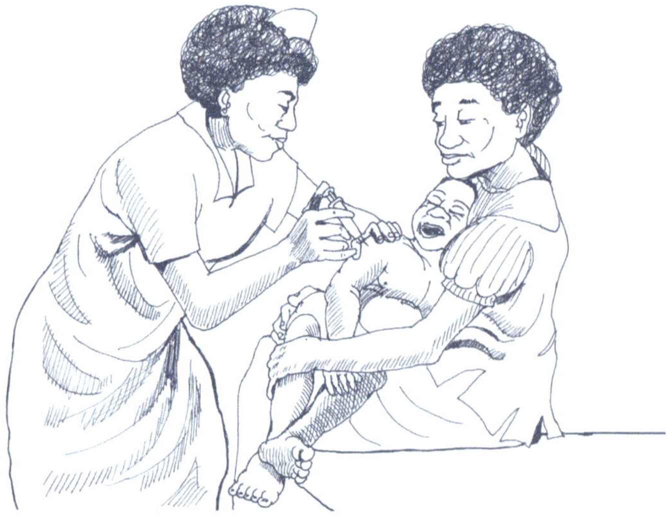

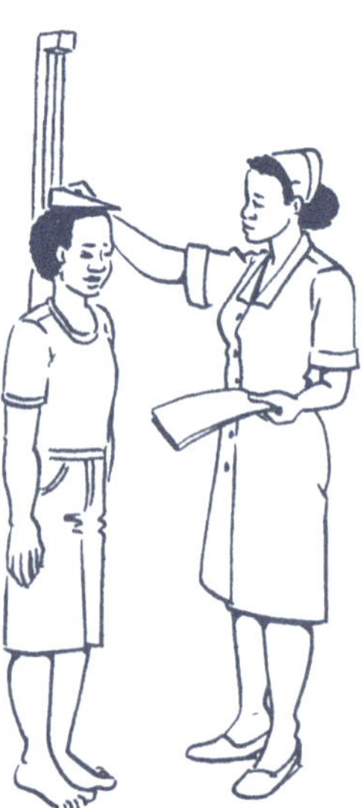

This table shows the maximum daily temperature recorded by some trekkers on the Kokoda Track during the dry season.

Maximum temperature (°C)	Frequency
24	1
25	2
26	5
27	6
28	3
29	2
30	3
31	1
32	1

1 a Is the data continuous or discrete?

b Construct an appropriate histogram or column graph of the data.

Data about the different illnesses diagnosed by a medical clinic were collected for a one-month period.

Illness diagnosed	Frequency
Measles	4
Malaria	8
Dysentery	7
Malnutrition	4
HIV/AIDS	1
Tuberculosis	1
Influenza	3

2 a Is the data continuous or discrete?

b Construct an appropriate histogram or column graph of the data.

This is a list of the weights (in kg) of students attending an elementary school.

17, 21, 15, 18, 19, 22, 23, 19, 17, 19, 21, 20, 21, 19, 18, 17, 18, 20, 21, 22, 19, 18, 17, 19, 18, 16

3 Present the data in:

a a frequency table including a relative frequency column

b an appropriate histogram or column graph

4 Using the data in question 3, find:

a the mean of the students' weights

b the mode of the students' weights

c the median of the students' weights

Help Box

In the last lesson you saw that data covering a wide range is better grouped into classes for graphing purposes.

Weight (kg)	Frequency
60–69	4
70–79	6
80–89	5
90–99	3

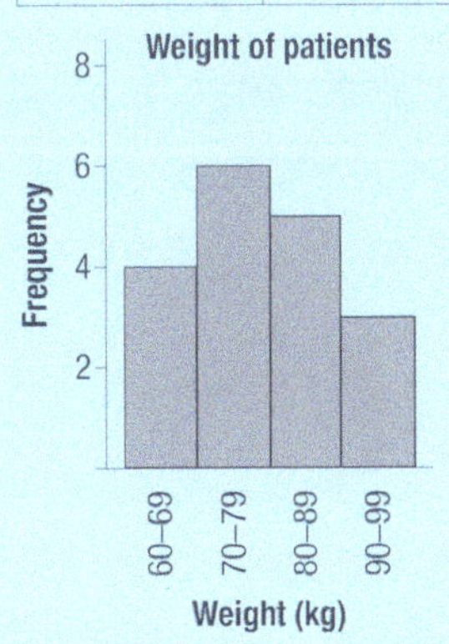

5 Group the following weight data into a frequency table using the classes 20–29, 30–39, 40–49, 50–59, 60–69, 70–79.

47, 33, 27, 58, 63, 71, 54, 39, 27, 24, 35, 50, 62,
48, 51, 29, 28, 60, 43, 36, 25, 33, 77, 69, 41, 56

6 Construct a histogram to display the data in question 5.

7 For the following frequency tables, state whether a histogram or a column graph should be used and draw the appropriate graph.

a Some footballers record their running speed over a certain distance.

Time (s)	40–49	50–59	60–69	70–79
Frequency	3	7	15	2

b A tally is kept of the age of patients attending a special clinic for those suffering skin infections.

Age	0–5	6–10	11–15	16–20	21–25	26–30
Frequency	5	12	8	4	2	5

c The main colour of the bilums brought to school.

Colour	Red	Yellow	Green	Blue
Frequency	8	2	4	7

Challenge

Collect two different sets of data relating to your class: one set that is appropriate to present in a histogram and the other set that is appropriate to present in a column graph. Construct a frequency table and matching graph for each set of data. Write a few sentences to explain how you decided which data was most suited to that type of graph.

Lesson 8 Who belongs where?

In Grade 5 you began learning about sets and Venn diagrams.

This Venn diagram classifies a group of infants into one or more sets depending on whether they have been immunised against three particular diseases.

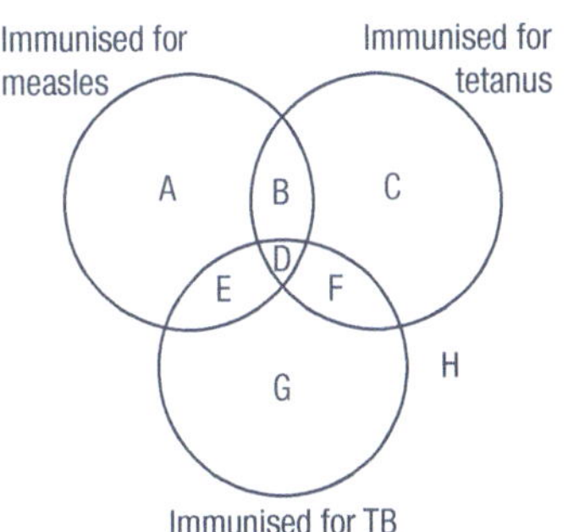

1 Define the attributes of the infants in each set from A to H.

2 The words in the left-hand column relate to sets. Copy the words into your workbook and match each with the most suitable descriptor from the right-hand column.

Subset	One 4-year-old male immunised against TB
Intersecting set	Females under 5 years immunised against tetanus
Element	Infants other than males or females under 5 years immunised against TB
Null set	Infants under 5 years immunised against measles and TB

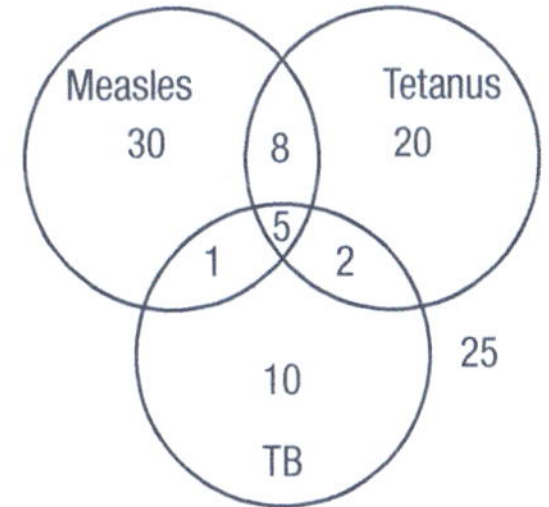

3 Answer these questions in relation to the above Venn diagram.

a How many infants under 5 years were immunised against tetanus only?

b How many infants under 5 years were immunised against TB only?

c How many infants under 5 years were immunised against both measles and tetanus?

d How many infants under 5 years are in the village altogether?

e How many infants under 5 years were immunised against all three diseases?

4 Twenty patients need to be immunised. Seventeen must be immunised against diphtheria and 11 against measles. Use a Venn diagram to show how many patients will receive both injections.

5 In a class of 25 students, 17 like reading, 9 like mathematics and three like neither. Use a Venn diagram to determine how many students:

a have a preference for mathematics

b have a preference for reading

6 Collect data and draw a Venn diagram to show how many of your class like reading, how many like mathematics and how many like both.

Challenge

Classify the numbers 1–10 into a Venn diagram to show the set of odd numbers and the set of prime numbers. What percentage of the ten numbers is in each set?

7 Use a Venn diagram to classify these shapes into triangles, quadrilaterals and shapes containing right angles.

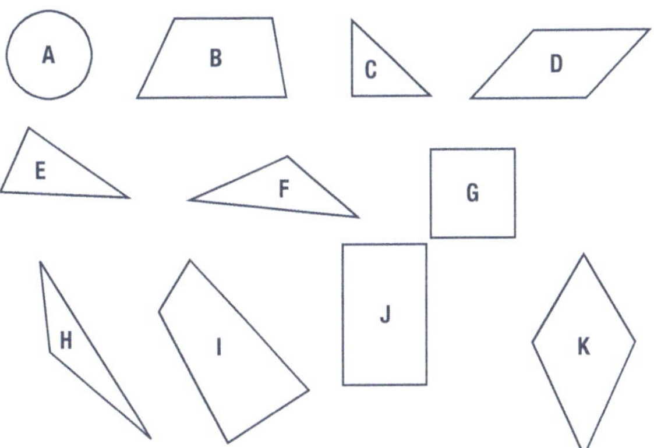

Lesson 9 Number patterns at work

Formulae are shortcuts that save people having to write a set of instructions in words to help solve problems. In earlier years you learned the formula for finding the area of a rectangle was $A = l \times w$ and for the area of a circle was $A = \pi r^2$.

Some people use the simple formula $F = 2C + 30$ to approximate the conversion of temperatures from degrees Celsius to degrees Fahrenheit. Using the formula is easier than writing 'To convert Celsius temperatures to Fahrenheit, double the degrees Celsius and then add thirty'.

1 Use the formula $F = 2C + 30$ to approximately convert the following Celsius temperatures to Fahrenheit temperatures.

a 17°C b 10°C c 19°C
d 32°C e 7°C f 25°C

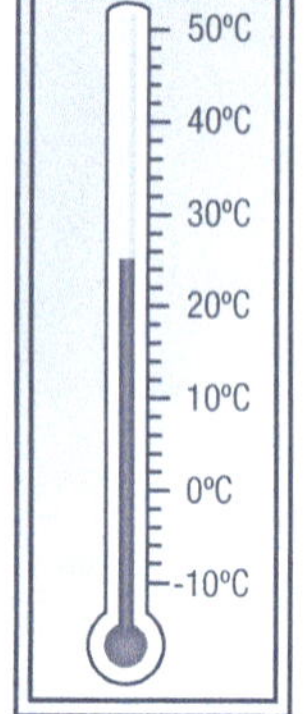

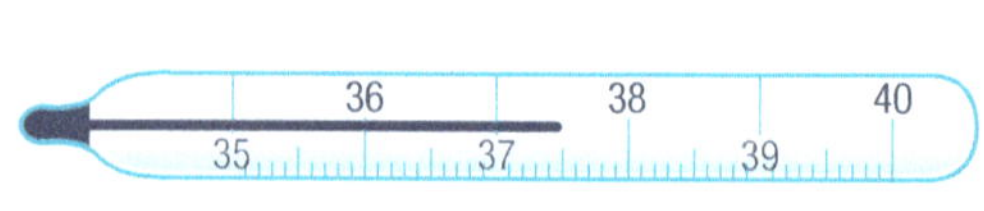

Challenge

Using the formula $F = 2C + 30$, create a formula that could be used to approximately convert Fahrenheit temperatures to Celsius temperatures.

A more accurate formula for converting Celsius temperatures to Fahrenheit is $F = \frac{9C}{5} + 32$.

2 Copy the following statement into your workbook and complete it to write the full explanation of the formula.

'To convert Celsius to Fahrenheit, multiply the Celsius temperature by _____, then divide by _____ and then add _____.'

3 Convert the following Celsius temperatures to Fahrenheit using the more accurate formula.

a 15°C b 22°C c 16°C d 34°C e 27°C f 11°C

4 Copy the following table into your workbook and complete it by using both formulae to convert Celsius to Fahrenheit.

Formula	9°C	13°C	18°C	26°C	29°C	34°C
$F = 2C + 30$						
$F = \frac{9C}{5} + 32$						

5 a Calculate the difference between each pair of answers obtained using the two formulae.

b Describe any patterns that you noted. For example, how consistently did one formula produce the lower value? Did the difference between the pairs of answers become wider or smaller as the temperatures increased?

In Grade 7 you learned about using diagrams and number patterns to help solve problems.

Help Box

Two nurses must be on duty in a particular hospital ward between the hours of 9 a.m. and 6 p.m. There are four nurses available. The dot and line diagram shows how many combinations of the four nurses are possible.

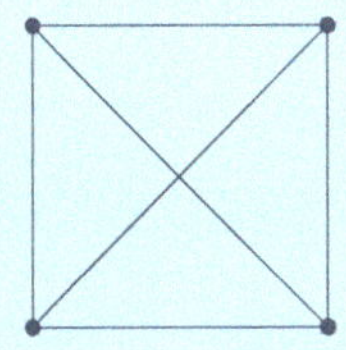

Four nurses can be organised into pairs six different ways.

6 Draw a dot and line diagram to help work out how many ways three nurses can be organised into pairs for the hospital roster.

7 Use diagrams to help work out how many ways the following numbers of nurses could be paired for the roster:

a two nurses b five nurses c six nurses d seven nurses

8 Copy and complete this table of values to show the possible combinations for different numbers of nurses.

Nurses	2	3	4	5	6	7	8	9
Paired combinations possible								

9 Use the completed table of values to answer these questions.

a Eight nurses are available to work in pairs in the ward. If a different combination of nurses is used each day, how many days will it take before each nurse has worked with each of the other seven nurses?

b If five nurses work in pairs and each nurse works only one shift with each of the other nurses, how many shifts will be involved?

c Nine new nurses are rostered to work in the ward in pairs. Each new nurse will work one shift only with each of the other new nurses. How many shifts will this involve altogether?

d Seven nurses were awarded a pay-rise. They all shake hands with each other once. How many handshakes were there?

Help Box

Using a table of values is more efficient than drawing dot and line diagrams. To calculate the values in the table we need to find the relationship between the items.

For example, to find the number of nursing pairs possible, you multiply the number of nurses by 1 less than the number of nurses and then halve it. So we can write a formula $\frac{n \times (n-1)}{2}$

So six nurses could be paired $\frac{6 \times (6-1)}{2} = \frac{6 \times 5}{2}$ which is 15 different ways.

10 Use the rule to find the number of ways that the following number of nurses could be paired.

a 18 nurses b 25 nurses c 29 nurses d 30 nurses

Learning Unit 3 Growing and Changing

Strand: Number and Application

Fractions	Outcome 8.1.1	Apply fractions in problem solving
Decimals	Outcome 8.1.2	Use decimals to solve real life problems
Decimals and Percentages	Outcome 8.1.4	Solve problems in any situation that involves percentages
Indices	Outcome 8.1.8	Use integer indices and fractional indices where the answers are rational

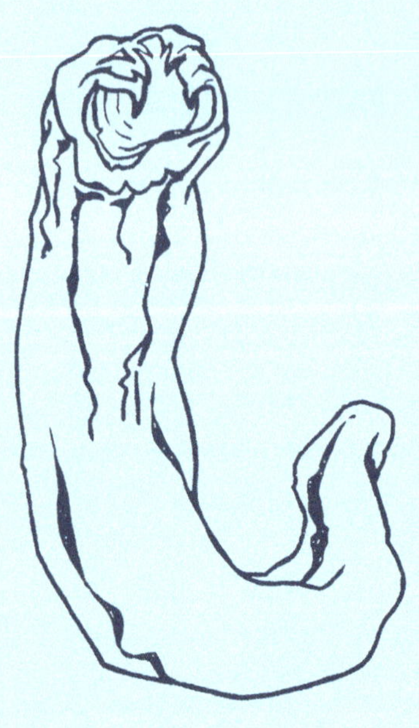

Lesson 1: Introduction	Plotting the potential population growth of different species
Lesson 2: A spoonful of medicine	Using decimals, fractions and percentages to solve problems
Lesson 3: Prime time	Creating factor trees Using prime factors in index notation Finding square and cube roots of numbers
Lesson 4: Back to basics	Recording numbers using index notation
Lesson 5: How many is that?	Identifying patterns formed by the use of indices Finding the square and cubic root of numbers
Lesson 6: Less than nothing	Calculating with zero as an exponent Working with negative exponents
Lesson 7: The spread of disease	Solving problems using exponents Using fractional exponents
Lesson 8: Health check	Applying formulae Calculating percentages, decimals and fractions

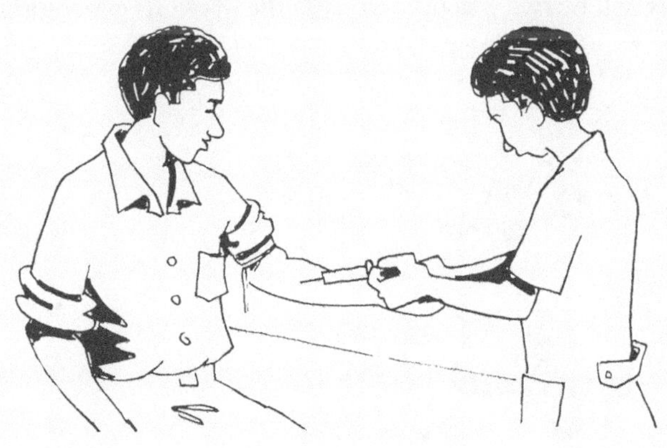

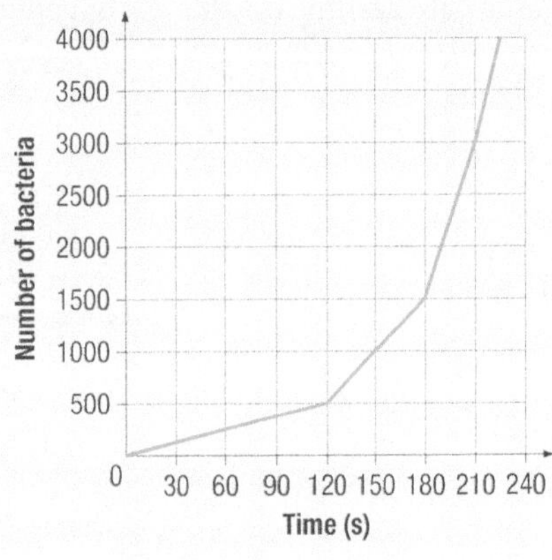

Lesson 1 Introduction

In this unit you will examine some of the different growth patterns in living organisms. The rapid growth rate of bacteria will be linked to the spread of disease, and the importance of preventative health care. The unit will build on your previous knowledge of number and extend your understanding of index notation.

Many organisms breed at a much faster rate than humans. The following table provides some comparisons of the potential increase in population of four different species over a two-year period.

Organism	Beginning population	Gestation period	Other relevant information	Estimated total population after 1 year	Estimated total population after 2 years
Human	2	9 months		2 + 1 = 3	3 + 1 = 4
Dog	2	9 weeks	Typically breed twice per year	2 + 2 = 4	4 + 2 = 6
Rabbit	2	4 weeks	Breed many times per year; typically 3 kittens per litter	2 + 18 = 20 (@ 6 litters)	20 + 18 = 38
Mouse	2	3 weeks	Breed many times per year; typically 10 kittens per litter	2 + 80 = 82 (@ 8 litters)	82 + 80 = 162

1 Draw a line graph of the data using a different coloured line to represent each species.

2 It is believed that mice can carry various diseases harmful to humans. Use the data in the table to write a few sentences explaining why mice have historically been credited with spreading epidemics and plagues.

3 Suggest reasons why mice rather than dogs are considered more likely to carry diseases harmful to humans.

4 List some other ways that you are aware of by which disease is spread.

5 Other than inoculation, what are some practical things people can do to prevent the spread of infection?

Bacteria are single celled organisms that increase in number by dividing themselves. This generally happens at a very rapid rate and means that each 'generation' of cells will be twice as numerous as the previous generation.

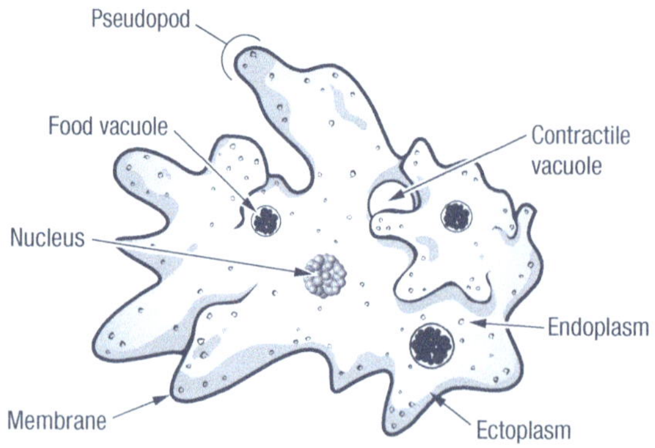

6 Draw a tree diagram to help you calculate the final size of the following bacteria population. One cell divided into two cells and these cells then each divided into two cells and this continued five more times.

Lesson 2 A spoonful of medicine

Medicine plays an important role in the health-care industry as a way of preventing disease, controlling the spread of infection, and as a cure for certain infections.

The data from a clinic in a major town showed that in one year a doctor had prescribed a total of 250 000 tablets to his patients for various illnesses.

1 Copy and complete the table below.

Type of tablet	Number	% of total
Vitamins		
Painkillers	83 275	
Antibiotics	42 805	
Antihistamines	58 680	

2 a If $\frac{2}{5}$ of the antibiotics were for skin infections, how many of the antibiotics were prescribed for other reasons?

b If $\frac{5}{8}$ of the vitamins were prescribed for women and children, how many vitamin tablets were prescribed for adult men?

c Antihistamine tablets are sold in packets of 36. How many packets did the clinic prescribe during the year?

d Forty per cent of the painkillers were prescribed for people between the ages of 20 and 40 years. How many tablets was that?

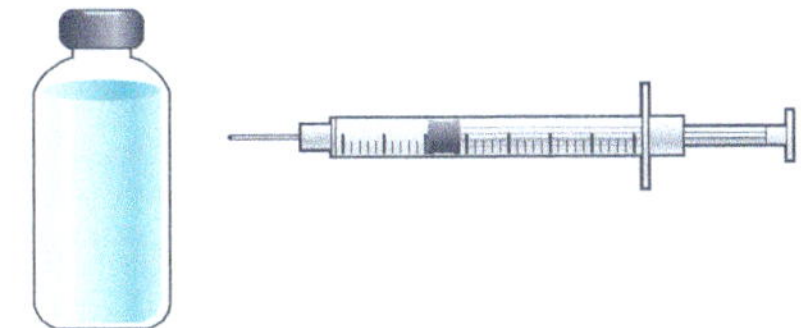

3 A particular vaccine comes in small bottles called phials. Each phial contains 0.7 mL. Calculate how many millilitres of vaccine are contained in:

a 10 phials b 50 phials

c 100 phials d 500 phials

e 750 phials f 1000 phials

4 Disposable syringes come in boxes of 100 and boxes of 500. Calculate the minimum number of each sized box that the clinic would need to order if they wanted:

a 350 needles b 1200 needles

c 780 needles d 3600 needles

5 A patient is given a 250 mL bottle of vitamin tonic.

a If the dose is 5 mL four times a day, how many whole days will the bottle last?

b How many mL of tonic will be left in the bottle after the last dose?

c How would the answers to parts a and b change if the medicine had to be taken three times daily instead of four?

6 The X-ray results from a certain clinic identified that 11 out of 55 patients had tuberculosis.

a What percentage of the patients was detected as having tuberculosis?

b If this trend continued, how many cases of tuberculosis would you expect to find after X-raying 350 patients?

7 The number of patients diagnosed with malaria at a village clinic increased by 5% from one year to the next. If 20 people were diagnosed with malaria during the first year, how many were diagnosed with malaria in the second year?

8 In a village of 460 people, 15% were inoculated against tuberculosis and 12 people were treated for malaria.

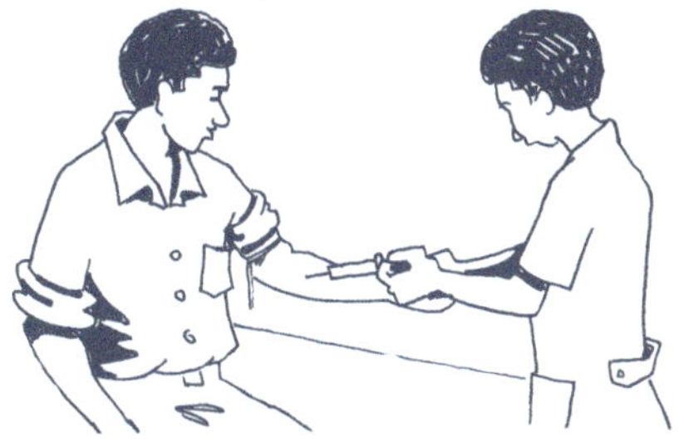

a How many people were inoculated against tuberculosis?

b What percentage of the village population was treated for malaria?

9 Six students in a class of 42 had measles and another seven had a vitamin deficiency.

a What fraction of the students had measles?

b What percentage of the students had a vitamin deficiency?

c What percentage of the students had neither measles nor a vitamin deficiency?

10 In one day a clinic distributed 120 antihistamine tablets, 360 painkillers and 80 antibiotic tablets.

a What fraction of the tablets listed are antibiotics?

b What percentage of the tablets listed were not painkillers?

c What was the ratio of antibiotics to antihistamines?

d If this was an average day at the clinic, how many days would it take until the clinic had distributed 25 000 painkillers?

Lesson 3 Prime time

When working with large numbers, it can be necessary to reduce each number to its prime factors. Understanding factors helps when simplifying and solving problems in algebra. This lesson will revise your knowledge of factors.

✻ ✻ ✻ ✻ ✻ ✻
✻ ✻ ✻ ✻ ✻ ✻
✻ ✻ ✻ ✻ ✻ ✻
✻ ✻ ✻ ✻ ✻ ✻

A group of 24 students organised themselves into six rows with four students in each row. The drawing on the left represents the students. Each row has an equal number of students, therefore 6 and 4 are **factors** of 24.

1 a Draw three other ways that the 24 students can be organised so that each row has an equal number of students.

b List the full set of factors of 24.

2 List the set of factors for each of the following numbers. Be systematic in your approach.

a 21 b 40 c 12 d 36
e 70 f 340 g 972

3 List the factors for each of the numbers from 1 to 20.

Help Box

The **factors** of a number are those numbers that can be divided into the number without leaving a remainder.

Some factors are a factor of more than one number. When two numbers share a factor, the factor is called a **common factor**.

4 a List the factors of 30.

b List the factors of 18.

c What factors appear in both lists and are therefore common factors of both 30 and 18?

5 Find the common factors of these numbers:

a 36, 58 b 12, 84 c 14, 49, 210
d 25, 45, 70 e 42, 126, 693

A **prime number** is a whole number that has only two factors: itself and 1. It can only be divided by itself and 1 without leaving a remainder. However the number 1 is not considered a prime number.

6 Use your answers from question 3 to help you list the prime numbers between 1 and 20.

Help Box

Each number can be written as a product of its prime factors. A factor tree can be used to find the prime factors of a number.

This example shows a factor tree for 780.

780
(2) × 390
(2) × 195
(5) × 39
(3) × (13)

$780 = 2 \times 2 \times 5 \times 3 \times 13$
$= 2^2 \times 5 \times 3 \times 13$

The prime factors of 780 are 2, 5, 3 and 13.

7 Use a factor tree to find the prime factors of the following numbers:

a 42 b 198 c 405 d 693

8 Find the prime factors and then all the factors of:

a 300 b 210 c 375 d 420

Challenge

List three numbers greater than 1000 that have 2, 3 and 5 as common factors. Use a factor tree to find the prime factors of each number.

Lesson  4 Back to basics

Earlier you learned that index notation is a handy way of writing some special number facts and of recording large numbers. Due to the large numbers involved in calculating the rapid rate of increase in bacteria, medical science is among the many professions that frequently use index notation in their calculations.

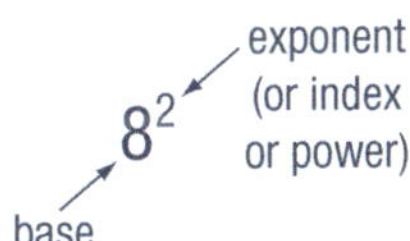

The terms **index**, **power** or **exponent** all mean the same thing. The index of a number shows how many times to use the base number in a multiplication.

$1^2 = 1$ and $2^2 = 4$; the numbers 1 and 4 begin the pattern of square numbers.

1 In order, list the square numbers from 1 to 100.

2 Write each of the following in index form.

a $3 \times 3 \times 3 \times 3 \times 3 \times 3 \times 3$

b $6 \times 6 \times 6 \times 6 \times 6$

c $10 \times 10 \times 10 \times 10$

d 4×4

e $m \times m \times m$

3 Write the following in expanded form. The first one has been done for you.

a $7^3 = 7 \times 7 \times 7$ b 9^2

c 12^4 d 10^2

e 11^5 f d^3

4 Write each of the following in index notation.

a 3 to the power of 4

b 8 squared

c 5 cubed

d three 2s multiplied together

e *m* to the power of 6

f nine 4s multiplied together

g *r* squared

h *b* to the fifth power

5 Calculate:

a 5^2 b 2^4 c 4^3 d 15^2

e 10^4 f 1^8 g 3^5

Exponent notation can be used to represent any number being multiplied by itself any number of times.

$$a^n = \underbrace{a \times a \times \ldots \times a}_{n}$$

6 Write the number in index form and find the value of the term a^n when:

a $a = 4, n = 2$ b $a = 6, n = 3$

c $a = 5, n = 4$ d $a = 1, n = 10$

7 Choose the correct answer.

a 4^3 written in factor form is:

A 4×3 B 3×4

C $3 \times 3 \times 3 \times 3$ D $4 \times 4 \times 4$

b x^2 written in factor form is:

A $x \times x$ B $2 \times x \times x$

C $2 \times 2 \times x$ D $2 \times 2 \times x \times x$

c $3(5^2)$ written in factor form is:

A $3 \times 5 \times 2$ B $3 \times 5 \times 5$

C $3 \times 3 \times 5 \times 5$ D $3 \times (5 \times 2)$

8 Which is the larger number in each pair?

a 9^2 or 2^9 b 2^4 or 4^2 c 10^3 or 3^{10}

d 4^3 or 3^4 e 6^2 or 2^6

Use the following examples to revise your understanding of brackets when the index number is the same for both factors and the base numbers are being multiplied together.

9 Calculate:

a $4^2 \times 9^2$ and $(4 \times 9)^2$

b $3^3 \times 2^3$ and $(3 \times 2)^3$

c $5^2 \times 4^2$ and $(5 \times 4)^2$

10 Use what you learned from question 9 to rewrite the following using brackets.

a $x^2 \times y^2$ b $y^3 \times b^3$ c $y^5 \times 3^5$

d $4^{10} \times x^{10}$ e $6^3 \times m^3 \times p^3$

11 Write the area of each of the following squares using index notation.

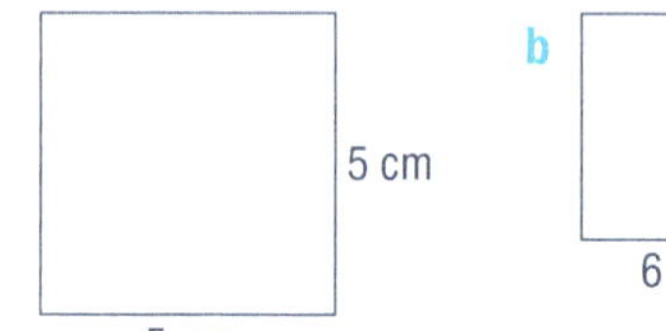

12 Write the volume of each cube using index notation.

a

2 cm

2 cm

2 cm

b

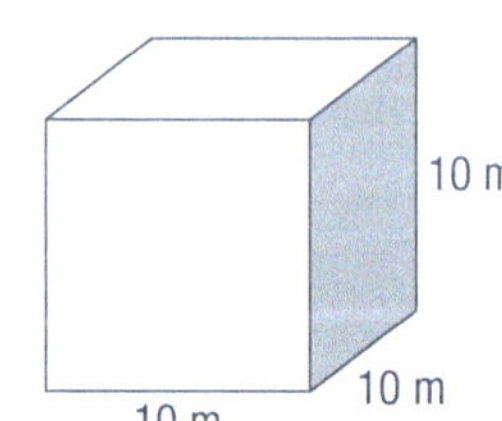

Challenge

Write an explanation for why a number 'to the power of two' is also known as a number 'squared' and a number 'to the power of three' is also known as a number 'cubed'.

Lesson 5 How many is that?

PUT YOUR SHOES ON AND BE WELL!

Papua New Guinea has more children infected with hookworm than many other places. Hookworm usually causes diarrhoea or cramps but can also cause tiredness, breathing difficulty, and heart problems. Infection can be serious for newborns, children, pregnant women, and people who are malnourished.

Infection can come from direct contact with contaminated soil, like walking barefoot.

SO PUT YOUR SHOES ON AND BE WELL!

A message from your health clinic.

A village nurse wanted to warn everyone about the dangers of hookworm. She gave one of these flyers to every patient she saw during one day, plus she gave them another two flyers and asked them to give a flyer to two other people.

1 a If the nurse saw 10 patients in the day, who each did as they were asked, how many people altogether should receive flyers?

b Which of the following descriptions best describes this number?

A $10 + 10$ B 10^2

C $10 + (10 \times 2)$ D 10^3

E $10 + 10^2$

Challenge

Write a story involving one of the incorrect statements from question 1 to show how the flyer would need to have been distributed for that statement to be correct.

The nurse also gave out some flyers at the village market. She gave one to her friend, three to a group of teenage boys, five to a group of mothers and seven to stallholders. Later when working out the total number of brochures she had given out at the market, the nurse noticed a pattern in the numbers. The pattern looked like this:

Number of brochures

$$1 \quad (1^2)$$
$$1 + 3 = 4 \quad (2^2)$$
$$1 + 3 + 5 = 9 \quad (3^2)$$
$$1 + 3 + 5 + 7 = 16 \quad (4^2)$$

2 a Copy and continue the pattern in your workbook to show the next three lines in the pattern.

b Write a few sentences to describe the pattern and explain how it could be used to work out the sum of the first 9 odd numbers without having to add them.

3 Use the pattern you described in question 2b to work out the sum of the first:

a 12 odd numbers

b 25 odd numbers

c 46 odd numbers

d 100 odd numbers

Challenge

Investigate whether a similar pattern involving index notation exists for calculating the sum of numerically ordered even numbers.

4 a Copy and complete these subtractions involving numerically ordered square numbers.

$2^2 - 1^2 = 4 - 1$ (or 3)

$3^2 - 2^2 = 9 - 4$ (or 5)

$4^2 - 3^2 = 16 - $ ___ (or ___)

$5^2 - 4^2 = $ _______ (or ___)

b Write the next three lines in the pattern.

5 In a few sentences, describe the pattern you noticed when doing question 4.

6 Use the pattern described above to solve the following without having to actually square the numbers.

a $12^2 - 11^2$ b $15^2 - 14^2$

c $473^2 - 472^2$ d $1082^2 - 1081^2$

7 a Use a calculator to find a pattern.

$1^2 = 1$

$11^2 = 121$

$111^2 = $ _____

$1111^2 = $ _____

b Write a few sentences to describe the pattern you noticed.

8 Without using a calculator, use the pattern described above to find the answers to:

a $111\,111^2$ b $11\,111^2$ c $1\,111\,111^2$

Help Box

The opposite of a square number is a square root. The symbol $\sqrt{\ }$ means square root.

So, the $\sqrt{25} = 5$ because 5×5 (or 5^2) $= 25$.

However, -5×5 (or -5^2) also equals 25.

So $\sqrt{25}$ = both 5 and –5.

Every positive number has two square roots.

9 List the square roots for:

a 16 b 64 c 81 d 100

e 121 f 144 g 225

Help Box

The opposite of a cube number is a cube root.

The symbol $\sqrt[3]{\ }$ means cube root.

So, the $\sqrt[3]{27} = 3$ because $3 \times 3 \times 3$ (or 3^3) $= 27$.

A number can only have one cube root.

10 Calculate:

a $\sqrt[3]{8}$ b $\sqrt[3]{27}$ c $\sqrt[3]{125}$

d $\sqrt[3]{343}$ e $\sqrt[3]{1000}$

11 Which number:

a multiplied by itself is equal to 36?

b has a cube root of 4?

c can be squared to make 100?

d has a square root of 11?

12 Which is the larger number in each pair?

a $\sqrt[3]{8}, 2^3$ b $\sqrt[3]{64}, -5^2$ c $\sqrt[3]{49}, 3^2$

d $\sqrt[3]{81}, 2^3$ e $3^5, -5^3$

Challenge

Use index notation to describe at least one number that is:

a between 5 and 28

b between 2 and 10

c between 18 and 30

d between 110 and 125

e between 150 and 170

f between 272 and 290

Lesson 6 Less than nothing

Until this lesson, you have only used indices larger than 1. However the numbers that can be used as indices extend well beyond that.

Help Box

If the exponent is 1, then the result is the number itself. For example, $7^1 = 7$.

If the exponent is 0, then the result is always 1. For example, $7^0 = 1$.

This is because exponent notation follows a pattern that involves the digit 1 in all calculations (even though this may not be written in many cases). The table shows the expanded version of the pattern.

Examples of powers of 4			
4^3	$1 \times 4 \times 4 \times 4$	64	↑ 4 times larger each time
4^2	$1 \times 4 \times 4$	16	
4^1	1×4	4	
4^0	1	1	

1 Copy the following table into your workbook and complete it.

$2^0 = 1$	$5^0 = 1$	$10^0 =$
$2^1 = 2$	$5^1 =$	$10^1 =$
$2^2 = 4$	$5^2 = 25$	$10^2 =$
$2^3 =$	$5^3 =$	$10^3 =$
$2^4 =$	$5^4 =$	$10^4 =$
$2^5 =$	$5^5 =$	$10^5 =$

2 Calculate:

a $3^1 + 3^0$ b $5^1 + 2^0$ c $2^2 + 6^1$
d $10^3 + 4^0$ e $5^2 + 12^0$ f $3^3 + 8^0$
g $12^1 + 36^0$ h $65^1 + 17^0$ i $24^1 + 3^3$
j $5^2 + 11^0$

Each month during 2008, a medical clinic noticed that the number of patients diagnosed as infected with hookworm tripled. In the first month there was 1, in the second month 3, in the third month 9 and so on each month.

3 Copy and complete the table using index notation to show the trend.

Month	1	2	3	4	5	6
Exponent	3^0	3^1	3^2			
Calculation	Always 1	1×3	$1 \times 3 \times 3$			
Monthly total	1	3				

4 Use a calculator to work out how many patients will present at the clinic with hookworm after:

a 9 months b 10 months c 12 months

Another clinic had one patient with hookworm in January. The number doubled exponentially every month after that for the next 12 months.

5 Create a table in your book to show the number of patients presenting with hookworm each month for the 12-month period.

Help Box

It is also possible to have negative exponents. While positive exponents use multiplication, negative exponents use division. A negative exponent means how many times **to divide** by the number.

For example, $5^1 = 1 \times 5 = 5$ $5^{-1} = 1 \div 5 = 0.2$

An easy method for remembering how to work with positive and negative exponents is to begin with 1 and then multiply or divide as many times as the exponent says.

6 Copy and complete the following equations. The first one has been done to help you.

a $2^{-3} = 1 \div 2 \div 2 \div 2 = 0.125$

b $2^{-5} = 1 \div 2 \div 2 \div 2 \div 2 \div 2 =$ _____

c $5^{-3} = 1 \div$ ____________ $=$ _____

d $10^{-2} = 1 \div$ ____________ $=$ _____

e $4^{-1} =$ __________ $=$ _____

f $3^{-2} = 1 \div$ ____________ $=$ _____

Challenge

Describe how you could check your answers to question 6 by using your understanding of the inverse relationship between division and multiplication.

7 Calculate the value of:

a	4^2	b	10^{-4}	c	2^0	d	1^{-2}
e	5^4	f	4^{-2}	g	10^2	h	100^{-2}
i	1^{10}	j	7^2	k	2^{-2}	l	3^{-3}

This table builds on the earlier table and shows the continuation of the pattern from positive to negative exponents.

Examples of powers of 4			
4^3	$1 \times 4 \times 4 \times 4$	64	↑ 4 times larger each time
4^2	$1 \times 4 \times 4$	16	
4^1	1×4	4	
4^0	1	1	
4^{-1}	$1 \div 4$	0.25	↓ 4 times smaller each time
4^{-2}	$1 \div 4 \div 4$	0.0625	
4^{-3}	$1 \div 4 \div 4 \div 4$	0.015625	

Help Box

Consider the following example that shows an easier way to calculate negative exponents.

For example, $2^{-3} = 1 \div 2 \div 2 \div 2$
$= 0.125$

2^{-3} means 1 divided by 2 three successive times:
$1 \div (2 \times 2 \times 2) = 1 \div 8$
$= \frac{1}{8}$ or 0.125

Negative exponents can therefore be converted into a positive fraction.

$$a^{-n} = \frac{1}{a^n}$$

So
$2^{-3} = 1 \div 2 \div 2 \div 2$
$2^{-3} = 1 \div (2 \times 2 \times 2)$
$= \frac{1}{2^3}$
$= \frac{1}{8}$ or 0.125

8 Use the $a^{-n} = \frac{1}{a^n}$ formula to change the following to their equivalent using a positive exponent.

a	5^{-3}	b	8^{-1}	c	4^{-2}	d	10^{-3}
e	6^{-2}	f	4^{-3}	g	6^{-1}	h	5^{-2}
i	2^{-4}	j	3^{-2}	k	2^{-3}	l	7^{-2}
m	m^{-2}	n	5^{-p}	o	x^{-y}		

9 Re-write the following to their equivalent using a negative exponent.

a	$\frac{1}{2^3}$	b	$\frac{1}{6^2}$	c	$\frac{1}{5^{10}}$	d	$\frac{1}{4^3}$
e	$\frac{1}{10^4}$	f	$\frac{1}{8^2}$	g	$\frac{1}{5^3}$	h	$\frac{1}{9^4}$
i	$\frac{1}{7^2}$	j	$\frac{1}{4^8}$	k	$\frac{1}{r^2}$	l	$\frac{1}{s^3}$
m	$\frac{1}{p^4}$	n	$\frac{1}{7^m}$	o	$\frac{1}{x^y}$		

Challenge

Use a calculator to find the actual value of answers to question 9, parts a to j.

Lesson 7 The spread of disease

The population growth of a certain type of bacteria over time (in seconds) is shown in the graph below.

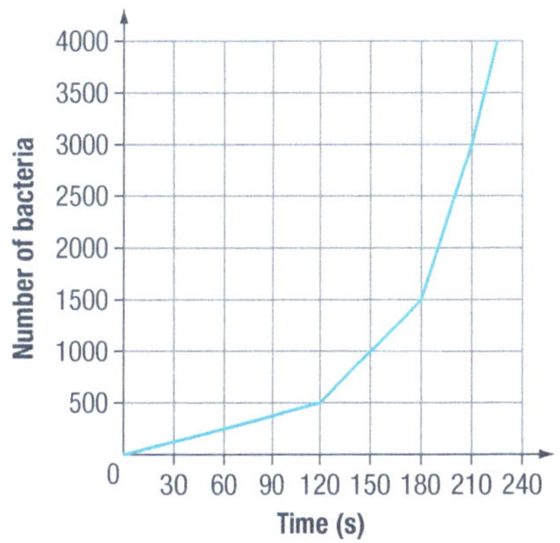

1 How many bacteria are there after:

a 1 minute b 2 minutes c 3 minutes d 3.5 minutes?

2 How long did it take the population of bacteria to reach:

a 1000 b 1500 c 3500?

3 A researcher said that each bacterial cell in his experiment divided at a rate of 10^4 every 12 hours. At that rate, how many cells would one cell have divided into in 12 hours?

4 A medication was found to destroy certain cancer cells at a rate of 10^{-3} over a 6-month period. What fraction of cells is this?

5 A researcher had collected 10 influenza cells in a Petri dish. Copy and complete the table to show how many cells there would be if they increased at each of the following rates over a 1-hour period.

Increase in cells at different growth rates		
Exponent form	**Number of cells**	**Sum of zeros in basic number**
10^1	10	1
10^2	100	2
10^3		
10^4		
10^5	100 000	
10^6		
10^7		7
10^8		
10^9		
10^{10}		

6 Copy and complete the following calculations by writing each number in factor form.

a $(3^2)^3 = (3 \times 3) \times (3 \times \_) \times (3 \times \_) = 3^6$

b $(2^3)^2 = (2 \times \_ \times \_) \times (\_ \times \_ \times \_) = 2^6$

c $(4^2)^2 = (\_ \times \_) \times (\_ \times \_) = \_\_\_$

d $(5^3)^3 = (\_ \times \_ \times \_) \times (\_ \times \_ \times \_) \times (\_ \times \_ \times \_) = \_\_\_$

e $(m^2)^3 = (\_ \times \_) \times (\_ \times \_) \times (\_ \times \_) = \_\_\_$

f $(p^2)^2 = (\_ \times \_) \times (\_ \times \_) = \_\_\_$

7 Look back at question 6. Write a few sentences to explain an easy way to calculate the answers to such calculations.

Help Box

It is possible for an exponent to be a fraction.

For example, $3^{\frac{1}{2}}$

When we square $3^{\frac{1}{2}}$ we find that $(3^{\frac{1}{2}})^2 = 3^{\frac{1}{2} \times 2}$
$= 3^1$
$= 3$

When we square $3^{\frac{1}{2}}$ we get 3, **so $3^{\frac{1}{2}}$ must be the square root of 3.**

8 Which of the following answers are correct?

a $5^{\frac{1}{2}}$ is the same as:

A $\frac{1}{2^5}$ B $\frac{1}{2^{-5}}$ C 5^{-2} D $\sqrt{5}$

b $\sqrt{3}$ is the same as:

A 3^2 B $\frac{1}{2^{-3}}$ C $3^{\frac{1}{2}}$ D 3^{-2}

c $4^{\frac{1}{2}}$ is the same as:

A $\sqrt{4}$ B $\frac{1}{2^{-4}}$ C 2^4 D 4^{-2}

d $\sqrt{100}$ is the same as:

A $10^{\frac{1}{2}}$ B $\frac{1}{2^{10}}$ C 10^2 D $\sqrt{10}$

9 Calculate the actual value of:

a $9^{\frac{1}{2}}$ b $4^{\frac{1}{2}}$ c $100^{\frac{1}{2}}$

d $m^{\frac{1}{2}}$ e $x^{\frac{1}{2}}$

Challenge

What is $27^{\frac{1}{3}}$?

Lesson  8 Health check

Earlier in this unit you learned hookworm is a problem in Papua New Guinea and that severe cases can cause breathing problems. In this lesson we will investigate some of the methods people use to check their breathing capacity.

The number of times a person breathes in one minute is known as their *respiration rate*. The average respiration rate for an adult at rest is 11 breaths per minute. During exercise an adult's respiration rate can increase to 35 breaths per minute.

1 Use a clock to work out your respiration rate while resting.

2 a Collect and present in a table, the respiration rate of at least five males and five females in your grade.

b Calculate the average respiration rates of each group.

3 Write a brief summary of what you found, being sure to answer the following questions.

a Is there a difference between male and female respiration rates?

b How does your respiration rate compare to the average rate for your gender?

c What is the range of the data?

d What were the highest and lowest rates recorded?

The average amount of air taken in (or expelled) is called the *breath volume*. The average breath volume for an adult resting is 0.6 litre. During exercise this can increase to 5 litres.

Help Box

The volume of air that a person breathes per minute is called their ventilation rate.

To calculate ventilation rate apply the formula:

ventilation rate = breath volume × respiration rate

For example, Harold is an adult with a respiration rate of 16 when resting.

So $0.6 \times 16 = 9.6$ L/min

Harold's ventilation rate is 9.6 litres per minute when resting.

4 Use the formula to calculate:

- a your ventilation rate when resting, assuming you have a breath volume of 0.5 L
- b the ventilation rate of an adult at rest with a breath volume of 0.7 L and a respiration rate of 17 breaths per min
- c a person walking with a breath volume of 1.3 L and a respiration rate of 24 breaths per min
- d a person jogging with a respiration rate of 33 breaths per min and a breath volume of 4.2 L

5 Look back at the information about Harold in the Help Box. Describe how the numbers in the calculation might alter if Harold had a serious case of hookworm and was having breathing problems.

The graph below shows the ventilation rate of a person before, during and after an exercise session. The period before exercise commenced is represented on the *x*-axis by negative numbers. The exercise session is shaded on the graph and begins on the *x*-axis at zero.

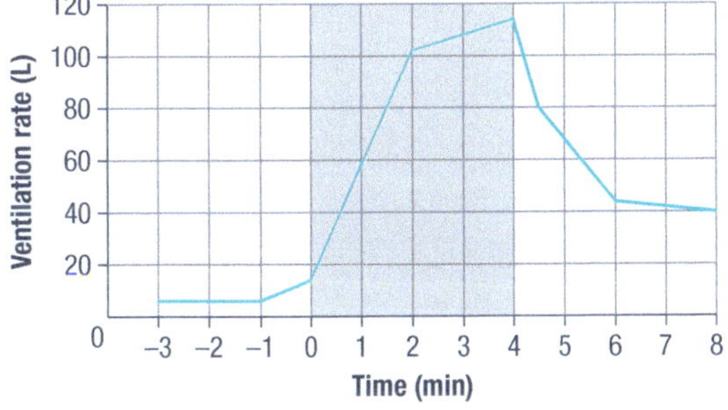

6
- a How long does the exercise session last?
- b What is the person's highest rate of ventilation?
- c How many minutes into the exercise session does the highest rate occur?
- d What is the difference between the person's ventilation rate before beginning the session and 4 minutes after the session ends?

7 Use the information on the graph to write a description of how the person's ventilation rate changes during the time covered.

8 A medical clinic visited a village school and found that 76% of the 125 students were infected with hookworm. Use your knowledge of percentage and fractions to answer the following questions.

- a 20% of the 125 students were males, how many were females?
- b What number of students was infected?
- c The clinic found that 15 of the infected students wore shoes all the time. What percentage of the infected students is this?
- d Two-fifths of all the students wore shoes to and from school and the others did not. How many students did not wear shoes when travelling to and from school?
- e The clinic found that 20 students had only a mild infection. What fraction of the whole school was that?

This table shows the percentage of public expenditure spent on health, education and defence in Papua New Guinea in two different years.

	Health	Education	Defence
1998	9.16%	16%	10%
1995	8.3%	16.2%	3.9%

9
- a What is the difference in the percentage spent on health between the two time-frames?
- b How much greater was the percentage spent on education in 1995 compared to the amount spent on health?
- c How does the amount spent on defence differ in the two years?
- d Write a few sentences to explain any changes you notice in spending priorities between the two years.

Challenge

Round the 1998 data in the table in question 9 to create a ratio of 9:16:10.

Write some statements showing what this means would be spent on each category if the total amount spent on the three items was K100, K5000, K800 000, K3 000 000.

Learning Unit  Additional Learning, Revision and Assessment

Strand: Number and Application

Indices	Outcome 8.1.8	Use integer indices and fractional indices where the answers are rational

Lesson 1: Additional Learning

Focus: Powers of 10

Lesson 2: Revision

Fractions, decimals, percentages and ratio
Weight
Temperature
Algebra
Statistics and graphs
Probability
Indices

Lesson 3: Assessment Task 1

Practical investigation—group work
Assessment Task 1 will assess learning outcomes 8.1.2 and 8.4.1.

40 marks

Lesson 4: Assessment Task 2

Test of basic skills and routine applications
Assessment Task 2 will cover the learning outcomes from across the topic 'A Lot Like Me'.
The test will assess the extent to which students can:

- Demonstrate an understanding of the mathematical concepts
- Correctly choose and apply mathematical techniques to solve problems

60 marks

Total 100 marks

Lesson 1 Additional learning

Powers of ten

During Learning Unit 3 you learned about index notation involving rational numbers. In this lesson you will discover the special value of using 10 as a base number for index notation. Using *powers of 10* is a very useful way of writing large numbers.

1 Copy and complete the following table.

Index form	Number	Zeros in the number
10^1	10	1
10^2		
10^3		
	100 000	5
10^6		
		7
10^{10}		

2 Write each of the following numbers in index form with a base of 10.

a one hundred

b ten thousand

c one million

d one hundred million

Help Box

Powers of 10 is commonly called **scientific notation** or **standard index form**.

In scientific notation a number is always written as $A \times 10^n$.

A is always a number between 1 and 10.

n is *to the power of.*

For example,
5 thousand is 5 times one thousand: $5000 = 5 \times 1000$

One thousand is 10^3

And 5 times 10^3 $= 5 \times 10^3$

$5000 = 5 \times 10^3$

3 Complete the following by writing each number using scientific notation.

a $400 = 4 \times 100$
$= 4 \times 10^{\_}$

b $7000 = 7 \times 1000$
$= 7 \times 10^{\_}$

c $80\,000 = 8 \times 10\,000$
$= 8 \times 10^{\_}$

d $9000 = 9 \times$ _____
$= 9 \times 10^{\_}$

e $300\,000 = 3 \times$ _____
$= 3 \times 10^{\_}$

f $6\,000\,000 = 6 \times$ _____
$= 6 \times 10^{\_}$

4 Copy and complete the following by writing each as a basic number.

a $3 \times 10^2 = 3 \times 100$
= _____

b $5 \times 10^3 = 5 \times 1000$
= _____

c $6 \times 10^4 = 6 \times$ _____
= _____

d $8 \times 10^5 = 8 \times$ _____
= _____

e $2 \times 10^7 = 2 \times$ _____
= _____

f $1 \times 10^6 = 1 \times$ _____
= _____

5 In one year, light travels at over 9×10^{15} metres. Write this distance as a basic number.

6 The mass of the sun is just under 2×10^{30} kg. Write this mass as a basic number.

7 The distance around the equator is about 40 000 km. Write this distance using scientific notation.

8 It is expected that the heat from the sun will make the Earth uninhabitable in about 55 million years. Write this amount using a power of ten.

Help Box

Scientific notation can also be used with decimal numbers.

The index of 10 shows how many places to move the decimal point to the right.

For example, $2.0 \times 10^4 = 2.0 \times (10 \times 10 \times 10 \times 10)$
$= 20\,000$

In the same way: $1.35 \times 10^4 = 1.35 \times (10 \times 10 \times 10 \times 10)$
$= 1.35 \times 10\,000$
$= 13\,500$

An easy method is to think 'move the decimal point 4 places to the right' like this:

1.35	→	13.5	→	135	→	1 350	→	13 500

9 The actual distance that light travels in one year is 9.46×10^{15} metres. Write this distance as a basic number.

10 The actual mass of the sun is 1.9891×10^{30} kg. Write this mass as a basic number.

11 Write the following as basic numbers.

a 1.5×10^7 b 3.5×10^3 c 1.85×10^5 d 7.56×10^5

e 3.6×10^{10} f 6.03×10^8 g 2.85×10^6 h 1.302×10^{10}

12 Copy and complete the following by writing in standard index form. (Remember $A \times 10^n$ and A must be between 1 and 10.)

a $15\,000\,000 = 1.5 \times 10^{-}$ b $38\,000 = 3.8 \times 10^{-}$ c $24\,500 = 2.45 \times 10^{-}$

d $260\,000 = 2.6 \times 10^{-}$ e $2\,875\,000 = 2.875 \times 10^{-}$ f $4\,570\,000 = 4.57 \times 10^{-}$

Challenge

How could you write the number 0.004 56 using scientific notation?

Lesson 2 Revision

Check your understanding of this topic using the following exercises.

Unit one

1 Answer the following questions using the fact that Mars has a gravitational pull that is about 0.38 of the Earth's.

- a If an object weighs 75 kg on Earth, what will it weigh on Mars?
- b If an object has a mass of 62.07 kg on Earth, what will be its mass be on Mars?

2 Copy and complete the table to show the estimated weight of the infants at each age if they increase their weight as per the column headings.

	Birth weight	Loss of 10% birth weight soon after birth	Increased birth weight by 30%	Increased birth weight by 50%	Double birth weight at 5 months
a	2.5 kg				
b	3200 g				
c	2.75 kg				

3 Calculate the weight range of the following newborns if they lost between 5% and 10% of their birth weight.

- a Birth weight 2700 g
- b Birth weight 2400 g
- c Birth weight 2380 g
- d Birth weight 1.7 kg
- e Birth weight 2.6 kg
- f Birth weight 2.05 kg

4 A baby's formula needs to be mixed using the ratio 1:3:5 (being 1 spoonful of tonic, 3 spoonfuls of milk powder and 5 spoonfuls of water).

- a How many spoonfuls of tonic are needed for 15 spoonfuls of water?
- b How many spoonfuls of milk powder are needed for 10 spoonfuls of water?
- c How many spoonfuls of water are needed for 6 spoonfuls of tonic?
- d How many spoonfuls of water are needed for 12 spoonfuls of milk powder?
- e What amount of each ingredient is contained in 63 spoonfuls of mixture?
- f How many spoonfuls of milk powder are contained in 27 spoonfuls of mixture?
- g If there were only 15 spoonfuls of milk powder remaining in the tin, how much mixture can be made using this ratio?

5 Calculate the weight of an adolescent who:

- a weighed 38.5 kg at age 13 and gained 6.75 kg during the next 2 years
- b began puberty weighing 29 kg and increased her weight by one-third
- c began puberty weighing 37 kg and increased his weight by 20%
- d weighed 46 kg at age 13 and increased her weight 15% in the first year of puberty and gained a further 5.35 kg in the second year of puberty

6 Find the BMI of the following females using the formula:

$$\text{BMI} = \frac{\text{weight in kilograms}}{\text{height in metres squared}}$$

a female 14 years old, height 1.2 m, weight 35 kg

b female 18 years old, height 1.6 m, weight 62 kg

c female 11 years old, height 150 cm, weight 34 kg

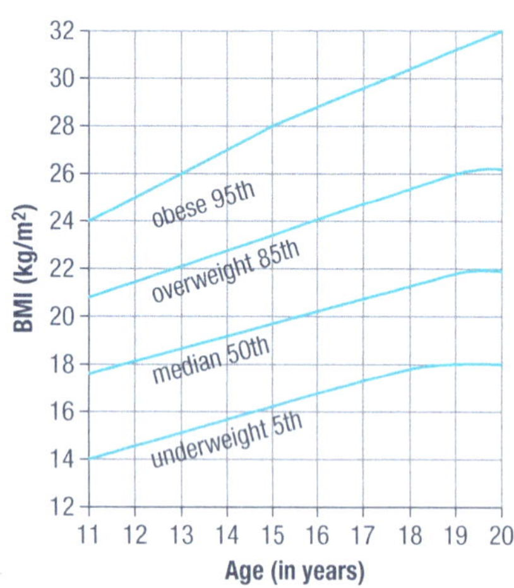

7 Use your answers from question 6 and the above graph to classify each of the females as underweight, normal, overweight or obese compared to an average female of that age. Explain your answers.

8 Copy and complete by showing the range of numbers covered by each of the following.

a 150 (± 10%) can be any answer within the range (______to ________)

b 160 (± 5%) can be

c 700 (± 5%) can be

d 65 (± 20%) can be

e 120 (± 25%) can be

9 Read the clinical thermometers and record the temperature to the nearest tenth of a degree.

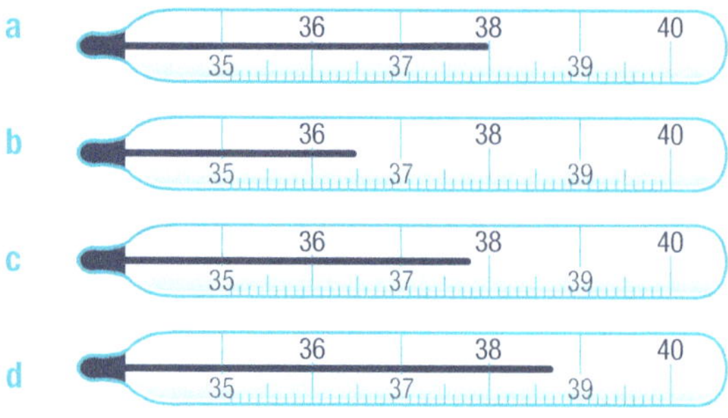

Unit two

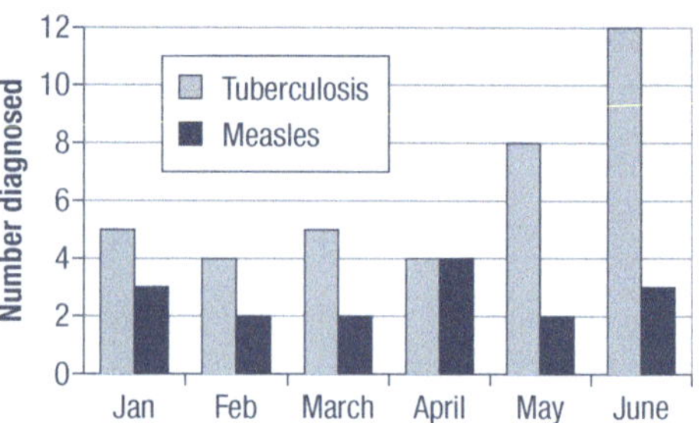

10 Use the data to answer the following questions.

a Which month had the highest number of measles cases diagnosed?

b What was the total number of patients diagnosed with tuberculosis during the 6-month period?

c Calculate the monthly mean of patients diagnosed with tuberculosis for the 6 months.

d Describe any trends that you notice in the data.

e Would it be reasonable to make the following statement based on this data: 'The major illness afflicting people in Papua New Guinea is tuberculosis'? Explain your answer.

11 The entire student population at a school of 300 students was immunised against malaria. A random sample of 50 students later found that two students had contracted the disease.

a What is the size of the population?

b What is the size of the sample?

c What fraction of the sample population did not contract malaria?

d Use this information to estimate what percentage of the entire student population may contract the disease.

12 Two friends wanted to improve their general health and decided to reduce the amount they each smoked. This table shows the number of smokes each friend had per week during a five-week period.

	Week 1	Week 2	Week 3	Week 4	Week 5
Ronnie	35	30	24	22	18
Amos	20	19	17	16	15

a Represent the data in a line graph using one line for each person.

b Calculate the mean number of cigarettes smoked by each of the friends for the five weeks.

c Write a few sentences to explain what trends you notice in the data.

d Predict what each man's data might look like in another five weeks if the current trends continue. Explain your predictions.

13 A large hospital collected data for six months about the age of children diagnosed with measles. The data is listed below:

5	7	5	8	11	10	6	7	4	5
4	5	11	12	7	9	4	5	11	6
10	5	9	5	6	4	5	7	6	11
7	10	6	9	8	6	4	10	5	4

a Create a frequency table of the data including a column showing relative frequency for each value.

b What was the most common age of the children in the study?

c What was the mean age of the children?

d What was the mode?

e What was the median age?

f What percentage of children in the study were under 8 years of age?

14 Write a few sentences to describe the difference between a column graph and a histogram. Include an example to show when it would be appropriate to use each.

15 Twenty-five patients need to be immunised. Fifteen must be immunised against malaria and 18 against tuberculosis. Use a Venn diagram to show how many patients will receive both injections.

16 Convert the following Celsius temperatures to Fahrenheit temperatures using the formula $F = \frac{9C}{5} + 32$

a 20°C b 32°C c 18°C

d 24°C e 14°C

Unit three

17 List the set of factors for each of the following numbers.

a 32 b 50 c 24 d 49

e 64 f 360 g 1024

18 Find the common factors of these numbers.

a 32, 40 b 12, 64 c 22, 33, 121

d 28, 42, 70 e 36, 72, 168

19 Use a factor tree to find the prime factors of the following numbers.

a 25 b 102 c 280 d 550

20 Which is the larger number in each pair?

a 3^2 or 2^3 b 5^4 or 4^5 c 10^2 or 2^{10}

d 5^3 or 3^5 e 1^2 or 2^1

21 Which of the following is the correct answer to $5^3 \times 4^3$?

A 20^6 B $(5 \times 4)^9$

C $(5 \times 4)^3$ D (15×12)

22 Which is the larger number in each pair?

a $\sqrt[3]{27}, 3^3$ b $\sqrt{64}, -5^3$ c $\sqrt[3]{125}, 2^2$

d $\sqrt{100}, 4^3$ e $2^5, -2^3$

23 Calculate:

a $4^1 + 2^0$ b $10^1 + 3^0$

c $5^2 + 2^3$ d $100^{-2} + 3^2$

24 Change the following to their equivalent using a positive exponent.

a 3^{-2} b 5^{-3} c y^{-2} d 5^{-g} e x^{-y}

25 Re-write the following to their equivalent using a negative exponent.

a $\frac{1}{6^2}$ b $\frac{1}{4^3}$ c $\frac{1}{h^2}$ d $\frac{1}{5^p}$ e $\frac{1}{x^y}$

26 A cancer cell was dividing at the rate of 10^6 per hour. How many cells would one cell have divided into after one hour?

27 A bacterial infection was dying at the rate of 10^{-2} per day. Express this in index notation form using a fraction.

Additional learning unit

28 Write the following as basic numbers.

a 3.5×10^4 b 6.13×10^5

c 1.75×10^3 d 3.06×10^{10}

29 Write in standard index form:

a $150\,000 = 1.5 \times 10^{-}$

b $260\,000\,000 =$ ____ $\times 10^{-}$

c $4500 =$ ____ $\times 10^{-}$

Assessment

Investigation

This assessment task gives you the chance to show what you understand about statistical data and the different ways that it is presented. The work you have done in the units *Different People, Different Sizes* and *The Past and the Future* will help you to complete this task.

Task 1: Practical investigation—group work

1 Collect five different types of graphs from newspapers, magazines and books. Name the type of graph used in each case, and write a paragraph describing the information displayed in each graph. If appropriate to the type of graph and the data presented, identify any trends and make predictions. (15 marks)

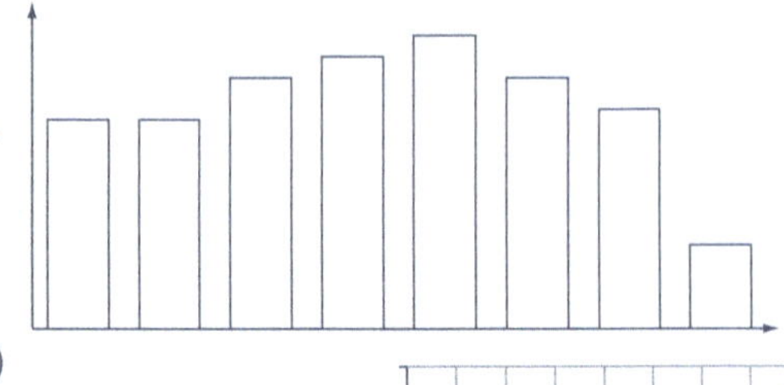

2 Collect data from at least ten people on a health issue of your choosing. The data must be able to be displayed in a histogram. Create a frequency table (including a relative frequency column) and draw the histogram to display the data collected. Write a list of ten questions (and their answers) that could be used by the teacher to test student's understanding of your frequency table and graph. (25 marks)

Your group will be assessed on your:

- ability to work productively as a group
- evidence of reasonable estimations made prior to calculations
- correct labelling of graphs
- ability to interpret data and make predictions
- correct presentation of a frequency table
- appropriate use of a histogram
- quality of questions posed to accompany histogram and frequency table
- correctness of calculations
- overall effort and persistence
- ability to complete the task in the time allocated.

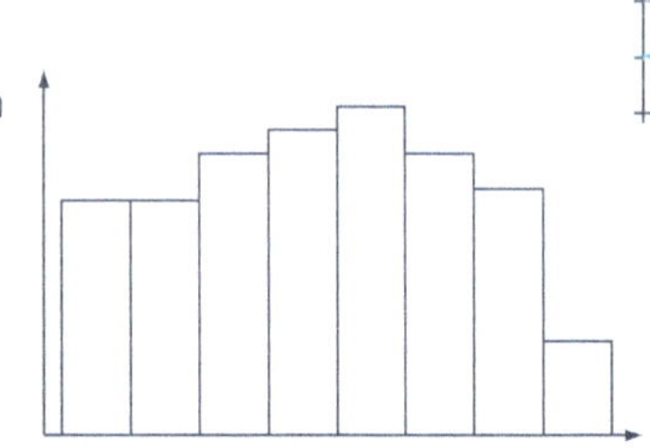

Total 40 marks

Assessment

Test

Task 2: Test of basic skills and routine applications

Mass and weight

1 Calculate the current weight range of these babies if they have lost 5–10% of their birth weight.

a birth weight 2.5 kg b birth weight 1900 g c birth weight 2250 g (3 marks)

2 Venus has a gravitational pull that is about 0.88 of the Earth's. A carton weighs 3.7 kg on Earth.

a What is the mass of the carton on Earth? b What would be the weight of the carton on Venus?

c What would be the mass of the carton on Venus? (3 marks)

Ratios

3 A recipe requires a mix in the ratio 1:3:4 (1 cup sugar, 3 cups flour, 4 cups water).

a How many cups of flour should be mixed with 4 cups of water?

b How many cups of sugar should be mixed with $4\frac{1}{2}$ cups of flour?

c How many cups of recipe mixture can be made if there is plenty of sugar and water but only one cup of flour left? (3 marks)

4 Calculate the range of acceptable answers for the following:

a 32 (± 2) b 3.5 (± 0.7) c 1.1 (± 0.05)

d 15 (± 20%) e 60 (± 15%) f 250 (± 5%) (6 marks)

5 Calculate the percentage error in the following estimations.

a estimated value 100, actual value 60 b estimated value 40, actual value 36

c estimated value 150, actual value 100 (3 marks)

Temperature

6 A dancer's body temperature was 36.4°C when she began dancing. Her temperature rose 0.4°C in the first 5 minutes, a further 0.2°C in the next 5 minutes and another 0.07°C before the dance ended. Five minutes later her temperature had fallen 0.3°C. What was her final body temperature? (1 mark)

Algebra

7 Use the formula $F = \frac{9C}{5} + 32$ to help you convert the following Celsius temperatures to Fahrenheit.

a 36°C b 26°C c 10°C d −3°C (2 marks)

8 Use the formula $M = \frac{n \times (n-1)}{2}$ to find the value of M when n equals:

a 9 b 15 c 23 d 29 (2 marks)

Percentage and fractions

9 The table shows the quantity of certain tablets dispensed by a chemist in one day.

a Calculate the percentage of each type of tablet listed in the table (to the nearest whole number).

b If $\frac{5}{6}$ of the antibiotics were for skin infections, how many of the antibiotics were prescribed for other reasons?

c Antihistamine tablets are sold in packets of 24. How many packets of antihistamine tablets did the chemist dispense on this day? (3 marks)

Tablets dispensed 2/4/08		
Type of tablet	Number	% of total
Vitamins	2400	
Painkillers	1344	
Antibiotics	288	
Antihistamines	480	

Assessment

Test

Statistics

10 These are the results of students who scored between 70 and 100 on a test. Draw a frequency table to represent the scores including a column showing relative frequency.

90, 72, 75, 83, 70, 90, 83, 77, 84, 72, 75, 70, 86, 86, 75, 89, 84, 72, 89, 76 (3 marks)

11 Answer the following questions using the data from question 10.

a How many of the scores are above 80?

b What percentage of the scores are less than 75?

c What is the mean score achieved by this group of students on this test?

d What was the median score? (4 marks)

12 a What does this graph show?

b From the graph, list the illnesses in order from most to least often diagnosed.

c Describe any trends in the graph.

d Predict what the hookworm data might be for the 2 weeks beyond this graph, if the current trend continues. Explain your answer.

e Would it be appropriate to make the following statement based on this data: 'The number of children being diagnosed with measles in Papua New Guinea is very small compared to other illnesses'? Explain your answer.

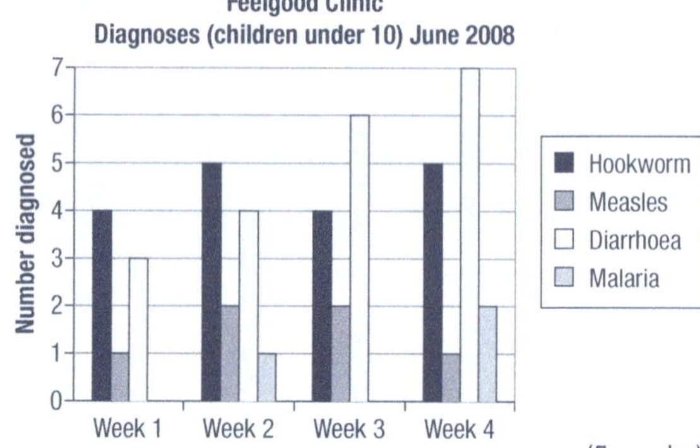

(5 marks)

13 a Group the following age data into a frequency table using the classes 10–19, 20–29, 30–39, 40–49, 50–59.

17, 23, 17, 18, 23, 51, 34, 39, 17, 24, 35, 40, 52, 38, 11, 19, 38, 30, 23, 24, 25, 23, 57, 49, 41, 36 (3 marks)

b Construct a histogram to display the data. (4 marks)

c Write a few sentences to describe the data in the histogram. (2 marks)

Indices

14 Calculate:

a $\sqrt[3]{27} + 5^2$ b $\sqrt{64} + 2^3$ c $5^3 \times 10^2$ d $10^1 + 3^0$ (2 marks)

15 Re-write the equivalent of the following using a positive exponent.

a 3^{-2} b 5^{-3} c 4^{-2} (3 marks)

16 Re-write the equivalent of the following using a negative exponent.

a $\frac{1}{8^2}$ b $\frac{1}{6^3}$ c $\frac{1}{5^6}$ (3 marks)

17 Write the following as basic numbers.

a 4.5×10^5 b 5.23×10^6 c 3.02×10^{10} (3 marks)

18 Write in standard index form:

a $250\,000 = \_\_\_ \times 10^{-}$ b $47\,000\,000 = \_\_\_ \times 10^{-}$ (2 marks)

Total 60 marks

Topic 4 The Tourism Market

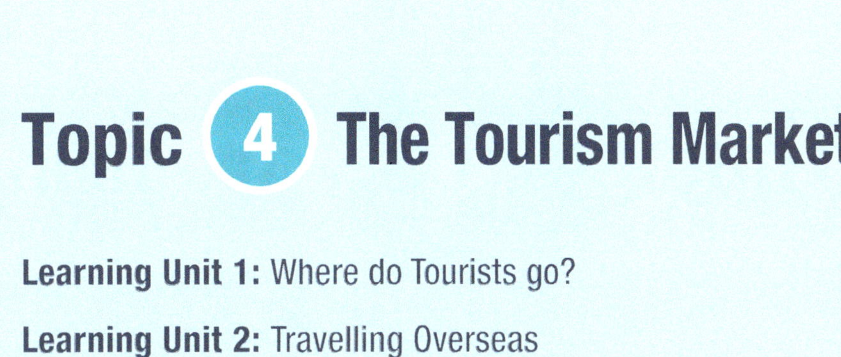

Learning Unit 1: Where do Tourists go?

Learning Unit 2: Travelling Overseas

Learning Unit 3: Tourism Comparisons

Learning Unit 4: Additional Learning, Revision and Assessment

Topic 4 The Tourism Market

In this topic you will investigate and compare some of the issues, facts and figures related to tourism both for people visiting Papua New Guinea and for people travelling to other countries.

The topic includes a closer look at the tourism industry in Papua New Guinea. We will also extend our horizons by looking at characteristics and features of other countries in the world and comparing them with Papua New Guinea.

The Assessment Tasks provide opportunity for you to demonstrate your understanding of the material covered both through project work and a formal test.

Topic 4: *The Tourism Market* comprises the following Learning Units:

Learning Unit 1: Where do Tourists go?

Learning Unit 2: Travelling Overseas

Learning Unit 3: Tourism Comparisons

Learning Unit 4: Additional Learning, Revision and Assessment

Overview

Each Learning Unit reflects an aspect of mathematics used in everyday life. A brief summary of the mathematics in each Learning Unit appears below.

Learning Unit 1, *Where do Tourists go?*, investigates scales when reading and drawing maps and applies rates to calculate problems associated with time, distance, speed and fuel consumption. It also extends previous work on fractions, decimals and percentages.

Learning Unit 2, *Travelling Overseas*, explores the mathematics involved with latitude and longitude, world temperature variations and time zones. Decimal calculations are required when solving problems about baggage weights, exchange rates and costs of airfares.

Learning Unit 3, *Tourism Comparisons*, requires decimal calculations to compare prices, heights and lengths of physical features in various countries. The unit also involves interpretation and presentation of some tourism statistics.

Learning Unit 4, *Additional Learning, Revision and Assessment*, includes a stand-alone lesson about pie charts, a revision lesson, an investigation and a formal test.

Some of the materials that you may need to complete the activities in this topic are listed below.

You may need:

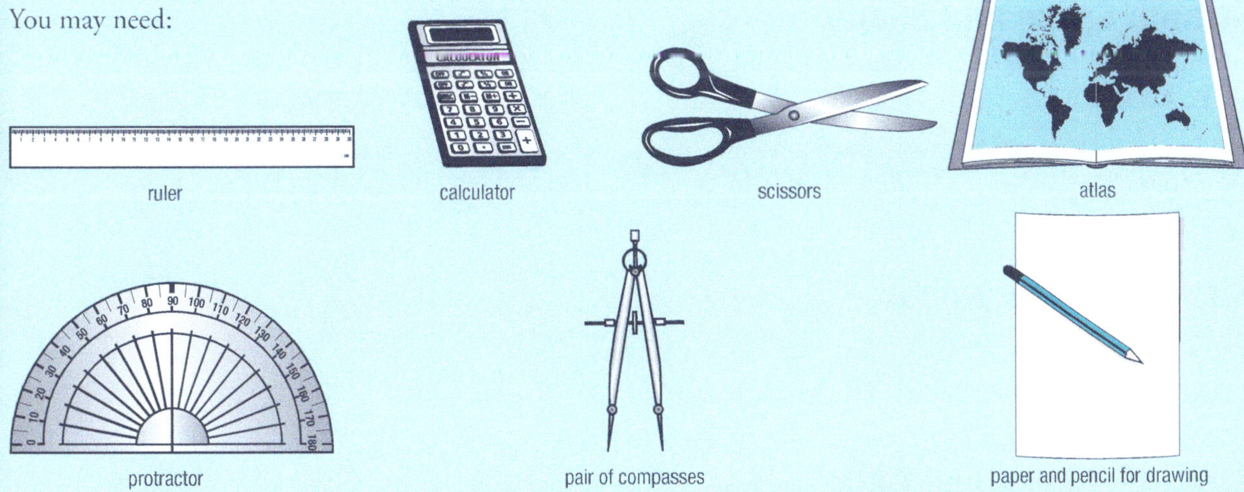

Learning Unit 1 Where do Tourists go?

Strand: Number and Application

Fractions	Outcome 8.1.1	Apply fractions in problem solving
Decimals	Outcome 8.1.2	Use decimals to solve real life problems
Fractions and Decimals	Outcome 8.1.3	Convert freely between fractions, decimals, percentages and ratios
Decimals and Percentages	Outcome 8.1.4	Solve problems in any situation that involves percentages
Ratios and Rates	Outcome 8.1.5	Apply ratios in solving problems from real life
Ratios and Rates	Outcome 8.1.6	Apply rates to solve simple problems from real life

Strand: Space and Shape

Tessellations	Outcome 8.2.10	Investigate rotational tessellations
Direction	Outcome 8.2.14	Use and read maps accurately

Strand: Measurement

Time	Outcome 8.3.5	Use time–rate calculations

Strand: Chance and Data

Estimation	Outcome 8.4.7	Estimate results of calculations

Lesson	Focus
Lesson 1: Introduction	Understanding the importance of tourism for Papua New Guinea
Lesson 2: Attractions of Papua New Guinea	Using a map key and scale Calculating distances on a map Converting between scale and ratio
Lesson 3: Flying in Papua New Guinea	Using direction to describe and locate places Using speed and time to determine distance travelled Using distance and speed to determine duration of flights
Lesson 4: The cost of airfares	Looking at the extra charges in airfares Calculating VAT (Value Added Tax) Calculating and estimating percentage discounts
Lesson 5: Where will we stay?	Calculating VAT applied in the ratio 1:10 Solving problems involving VAT Solving problems involving decimals and percentages
Lesson 6: What will we do?	Calculating VAT Solving problems with money
Lesson 7: Patterns in handcrafts	Identifying and drawing shapes that have rotational symmetry Creating tessellations that have rotational symmetry
Lesson 8: Renting a car	Estimating and calculating the amount of fuel used to cover specified distances Calculating fuel costs and fuel consumption
Lesson 9: Planning a journey	Constructing a sketch map Reading and constructing a scale map
Lesson 10: Holiday or work?	Writing tourism statistics as percentages Calculating percentage increases and decreases

Lesson 1 Introduction

Many people visiting Papua New Guinea are tourists who are interested in what it has to offer them. During this unit you will use the conventions of mapping to locate and specify places of interest to tourists, calculate distances between places, and construct a scale map. You will also investigate some of the issues and costs involved when tourists visit Papua New Guinea.

Tourism has the potential to be a very important industry for the people of Papua New Guinea. As well as being a major source of revenue, a healthy tourism industry can provide employment for many people.

Each picture below shows a situation or place where tourists might be found.

1 List some of the employment opportunities for locals that might exist in each situation.

2 List the ways tourists might spend their money in each situation.

Papua New Guinea's tourism industry is still very small. Factors hindering its growth include the lack of large enough airports to take wide-bodied jets, tribal fights, a high crime rate, rugged terrain, high prices for accommodation and travel within the country, and poor road networks to many areas.

3 Make a list of things that Papua New Guinea could do that might encourage more tourists to visit.

Lesson 2 Attractions of Papua New Guinea

Papua New Guinea has a huge range of spectacular and unique attractions. The map below shows some of the places and attractions that could interest tourists.

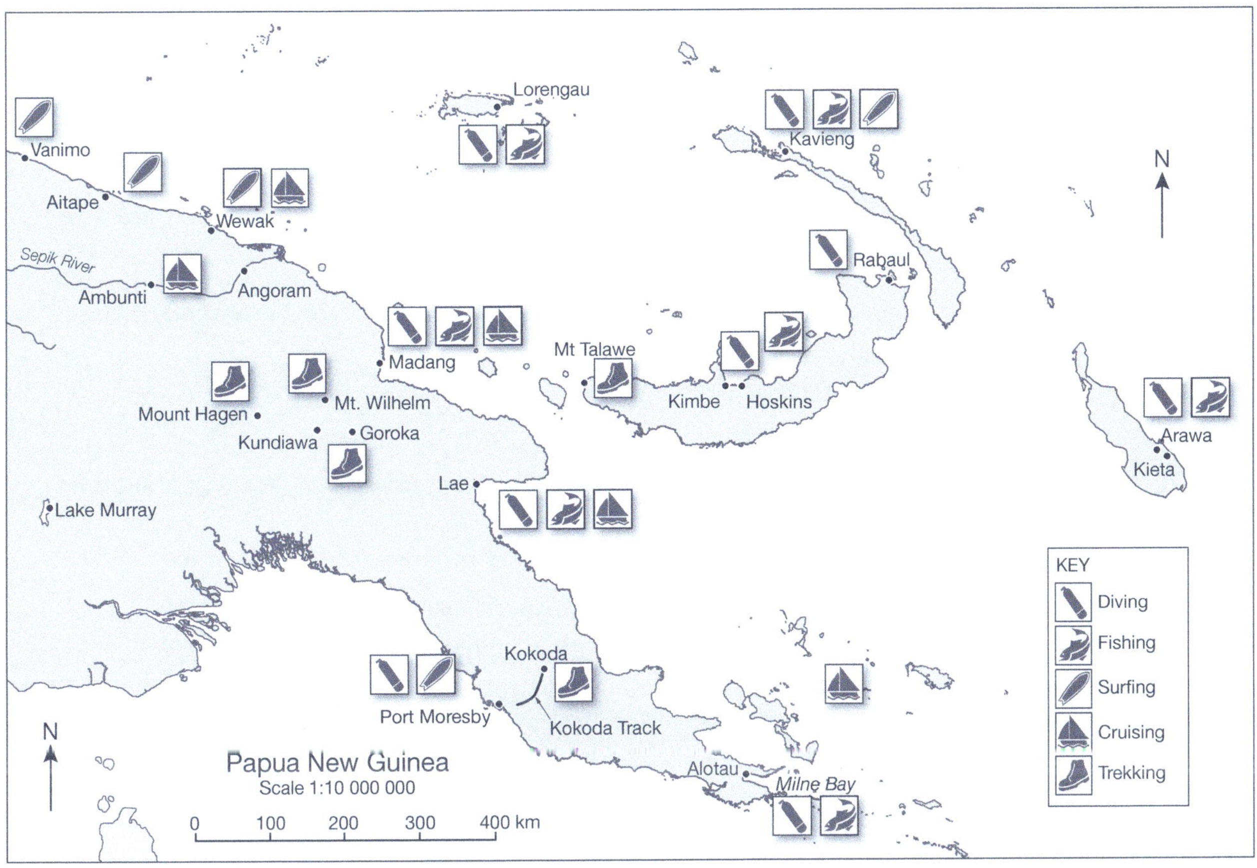

Use this map to help you answer questions 1 to 9.

1 You will see the scale 1:10 000 000 written on the map.

a What does this mean?

b How much smaller is this map of Papua New Guinea than the actual size of the country? Write your answer as a fraction.

2 Another map of Papua New Guinea has a scale of 1:20 000 000.

a Would it be larger or smaller than the map above?

b How much smaller would this map be than the actual country of Papua New Guinea? Write your answer as a fraction.

3 A map of Papua New Guinea has a scale of 1: 2 500 000.

a Would it be larger or smaller than the map on the previous page?

b How much smaller would this map be than the actual country of Papua New Guinea? Write your answer as a fraction.

4 How many kilometres along the ground in the real world is shown by one centimetre on the map?

5 What lengths on the map on the previous page would represent the following distances on the ground?

a	200 km	b	500 km	c	1200 km
d	850 km	e	450 km	f	325 km
g	275 km	h	920 km	i	1075 km
j	780 km				

6 What actual distances on the ground are represented by these measurements on the map on the previous page?

a	2.5 cm	b	4.2 cm	c	8.7 cm
d	5.8 cm	e	6.75 cm	f	10.25 cm
g	7.9 cm	h	3.1 cm	i	12.75 cm
j	1.25 cm				

7 Use a ruler and the map scale to find the approximate distances, if you travel in a straight line, between the following places:

a Port Moresby and Alotau

b Kimbe and Rabaul

c Wewak and Madang

d Goroka and Lae

e Mount Hagen and Port Moresby

f Lorengau and Madang

g Vanimo and Kavieng

h Alotau and Kieta

8 If a tourist planned to fly in a straight line from Port Moresby to Madang and then on to Alotau before returning to Port Moresby, approximately how many kilometres would they have travelled altogether?

9 Name two places on the map that are:

a about 200 km apart

b between 500 and 600 km apart

c more than 1000 km apart

Help Box

As we have seen, when writing a ratio as a scale it is important to remember that both amounts have the same unit of measurement.

For example, a map scale is shown as 1:8 000 000. This means that every 1 centimetre on the map represents 8 000 000 centimetres in actual distance.

When using a ruler to calculate distance on a map it is more useful to convert the centimetres to metres or kilometres, so 8 000 000 cm converted to kilometres becomes 80 kilometres.

The ratio 1:8 000 000 is the same as 1 cm = 80 km.

10 Convert the centimetres in these ratios to metres or kilometres.

a	1:15 000 000	b	1:275 000
c	1:50 000	d	1:100 000
e	1:6 500 000	f	1:7500

11 Convert both measurements in each scale to centimetres and then write the scale as a ratio.

a	1 cm = 50 km	b	1 cm = 3 m
c	1 cm = 12.5 km	d	1 cm = 50 m
e	1 cm = 25 km	f	1 cm = 150 m

Challenge

Using the scale 1:1500, draw a representation of a rectangular area that is 105 metres long and 75 metres wide.

Many tourists to Papua New Guinea are interested in diving, fishing, trekking, surfing or taking a cruise. The map of Papua New Guinea on page 69 has symbols to show you some of the places where tourists can participate in these activities.

12 Use the map key and the map of Papua New Guinea to help you answer these questions.

- a Name two places where tourists might go to surf.
- b What activities are available for tourists who visit Kavieng?
- c Where might tourists go if they want to go trekking?
- d Name some places where tourists can fish and dive.
- e What activities are available for tourists who want to stay within 150 km of Port Moresby?

Some tourists visit Papua New Guinea because they are interested in the volcanoes around Rabaul. The map below shows the position of the small volcanoes around Rabaul.

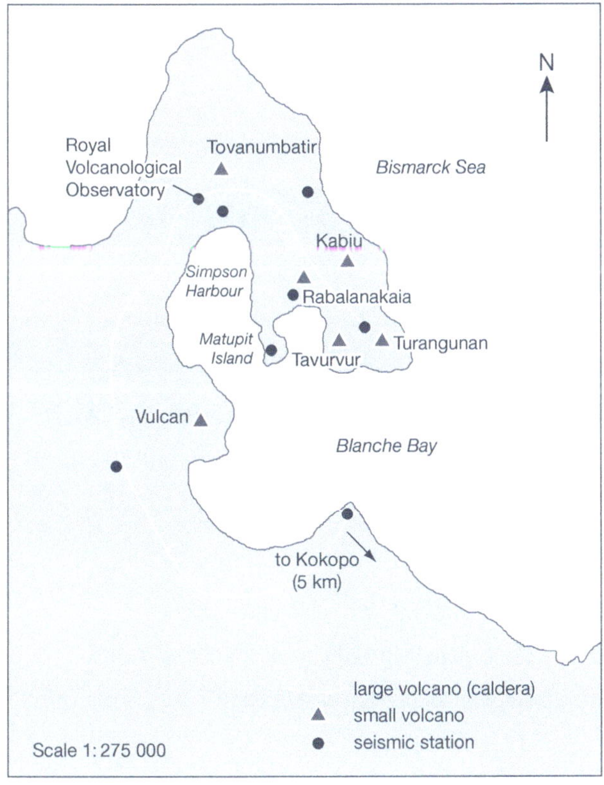

13 Answer these questions about the map.

- a How many small volcanoes are shown on the map?
- b How many seismic stations are shown?

14 Use the map scale to work out the approximate straight-line distances in kilometres between the following volcanoes:

- a Vulcan and Tavurvur
- b Tovanumbatir and Turangunan
- c Turangunan and Kabiu
- d Rabalanakaia and Vulcan
- e Vulcan and Tovanumbatir

A caldera is a large depression formed when magma erupts from beneath a volcano and the ground collapses into the emptied space.

15 Approximately how many kilometres is it from one side of the caldera to the other at the widest point in a north–south direction?

16 If you walked from the Royal Volcanological Observatory around the caldera to the most southerly seismic station and then continued your journey to Kokopo, about how far would you have walked?

Lesson 3 Flying in Papua New Guinea

Most tourists who come to Papua New Guinea fly into Jackson's Airport in Port Moresby, which is the only international airport. Once tourists are in Papua New Guinea, they can then travel by aeroplane to many places within the country. Some of these places are marked on the map below.

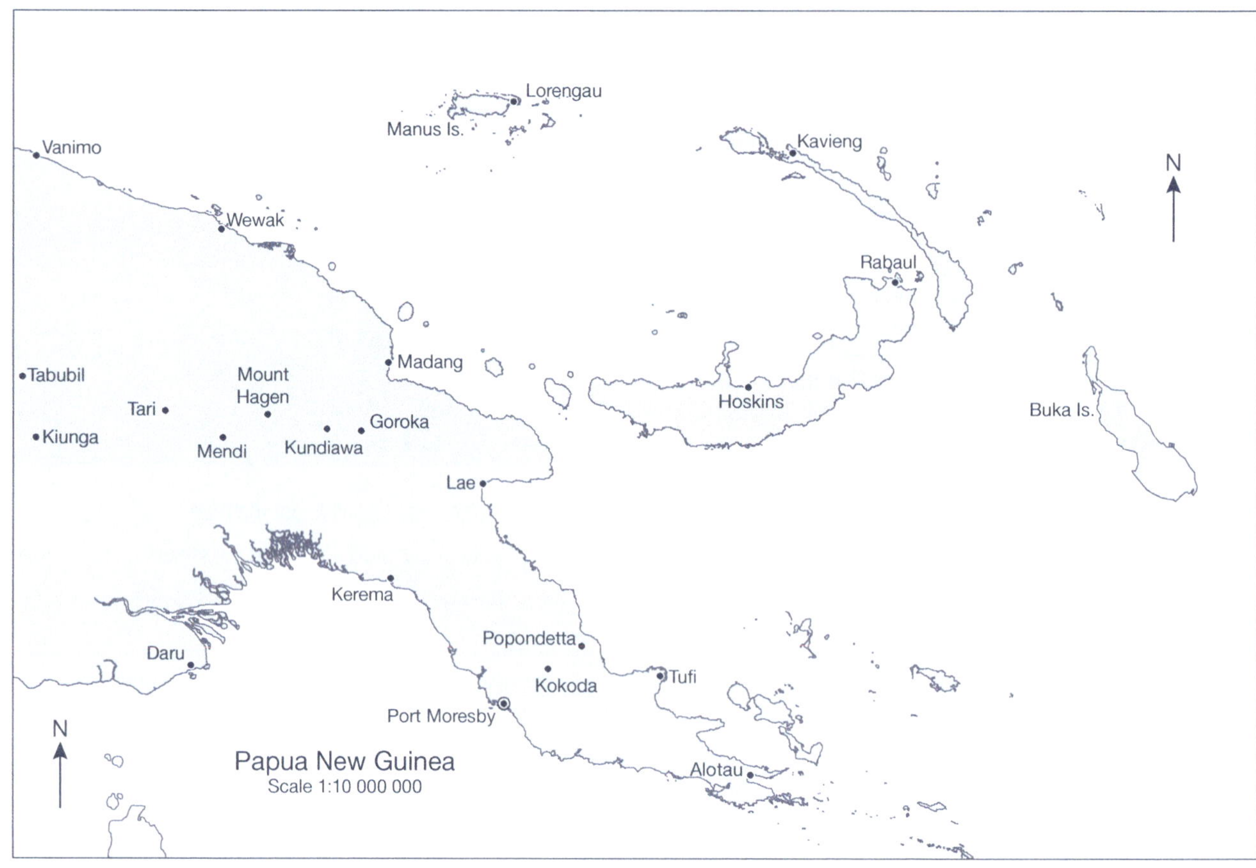

1 Use the map scale and your knowledge of direction to explain where the following places are in relation to Port Moresby. For example, Goroka is about 400 kilometres north-west of Port Moresby.

a Lorengau
b Rabaul
c Alotau
d Mendi

2 What are these places?

a A place about 225 km north-west of Port Moresby.
b A place about 600 km south of Wewak.
c A place about 800 km south-west of Kavieng.
d A place more than 1200 km south-east of Vanimo.
e A place about 550 km north-east of Kerema.
f A place about 650 km east of Daru.

Different types of planes travel at different speeds, so flights can take more or less time depending on the type of plane in which you travel.

Help Box

The speed of a plane can be used to calculate the distance travelled.

For example, a plane flies at a speed of 550 kilometres per hour (550 kph), and the flight takes 1 hour and 20 minutes.

This means that the plane covers a distance of 550 km every hour.

To calculate the distance travelled we can use the formula:

Distance = speed × time ($D = s \times t$)

Because 20 minutes is $\frac{1}{3}$ of an hour we can say

$550 \times 1\frac{1}{3}$, which is the same as $\frac{550}{1} \times \frac{4}{3}$,

so $\frac{550}{1} \times \frac{4}{3} = \frac{2200}{3}$

$= 733.33$

The distance travelled in 1 hour and 20 minutes was 733.33 kilometres.

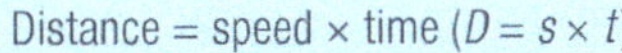

3 Use the method above to calculate the distance travelled by the following planes.

a A plane that flies at a speed of 480 kph and takes 45 min to reach its destination.

b A plane with a flying time of 2 h and 30 min and a speed of 230 kph.

c A plane that has a speed of 780 kph and reaches its destination in 40 min.

d A plane with a flying time of $\frac{3}{4}$ h and a speed of 300 kph.

e A plane that has a speed of 550 kph and takes 1 h and 15 min to reach its destination.

4 Use the above method and the map on the previous page to help you answer these questions.

a If a plane set out from Port Moresby and flew in a north easterly direction at 500 kph and landed approximately $1\frac{1}{2}$ hours later, where might it have landed?

b If a plane flies at 360 kph in a south easterly direction from Vanimo, where might it land $1\frac{1}{4}$ hours later?

c If a plane set out from Lae and travelled in a north westerly direction at 300 kph for 40 min, what part of Papua New Guinea would it be flying over?

d If you left Tufi and flew for 2 hours at a speed of 275 kph, where might you be in Papua New Guinea?

Help Box

If we know the speed of a plane and the distance travelled, we can calculate the flying time.

For example, we know a plane's speed is 550 kph and it is flying from Port Moresby to Rabaul, which is about 775 km.

To calculate the flying time we can use the formula:

Time = distance ÷ speed ($T = \frac{d}{s}$)

Therefore 775 ÷ 550 = 1.41 or 1 and $\frac{41}{100}$ hours.

To work out how many minutes $\frac{41}{100}$ represents, we say

0.41 × 60 (because there are 60 min in an hour).

This works out at 25 minutes (after rounding).

So 1.41 is the same as 1 h and 25 min.

The flying time between Port Moresby and Rabaul for a plane travelling at a speed of 550 kph is 1 hour and 25 minutes.

5 Use the method in the Help Box to calculate the flying time between Port Moresby and Rabaul for aircraft with a speed of:

a 780 kph b 500 kph c 230 kph

d 305 kph e 460 kph f 370 kph

6 Use the method in the Help Box and the map on page 72 to help you answer these questions.

a A plane flew from Wewak to Madang at a speed of 220 kph. What was its flying time?

b How long will it take to fly from Vanimo to Mendi at a speed of 370 kph?

c If we catch a plane that travels at a speed of 780 kph, what time will we reach Port Moresby if we leave Lorengau at 3 p.m.?

d We leave Kavieng at 10 a.m. on a plane travelling at 450 kph. What time will we arrive in Lae?

Challenge

Use what you have learned in this lesson to work out a formula for calculating plane speed when you know the distance travelled and the duration of the flight. Use your formula to work out what speed a plane is travelling if it takes 1.75 hours to travel between Daru and Buka Island.

Lesson 4 The cost of airfares

Airfares can change frequently so passengers need to check with airline offices or travel agents for up-to-date fares or for special fares that may be available.

The chart on the next page shows examples of adult airfares within Papua New Guinea for a one-way trip. The circled fares indicate that the flight is not direct, that is, to get from the place of departure to the destination the passenger has to take two flights.

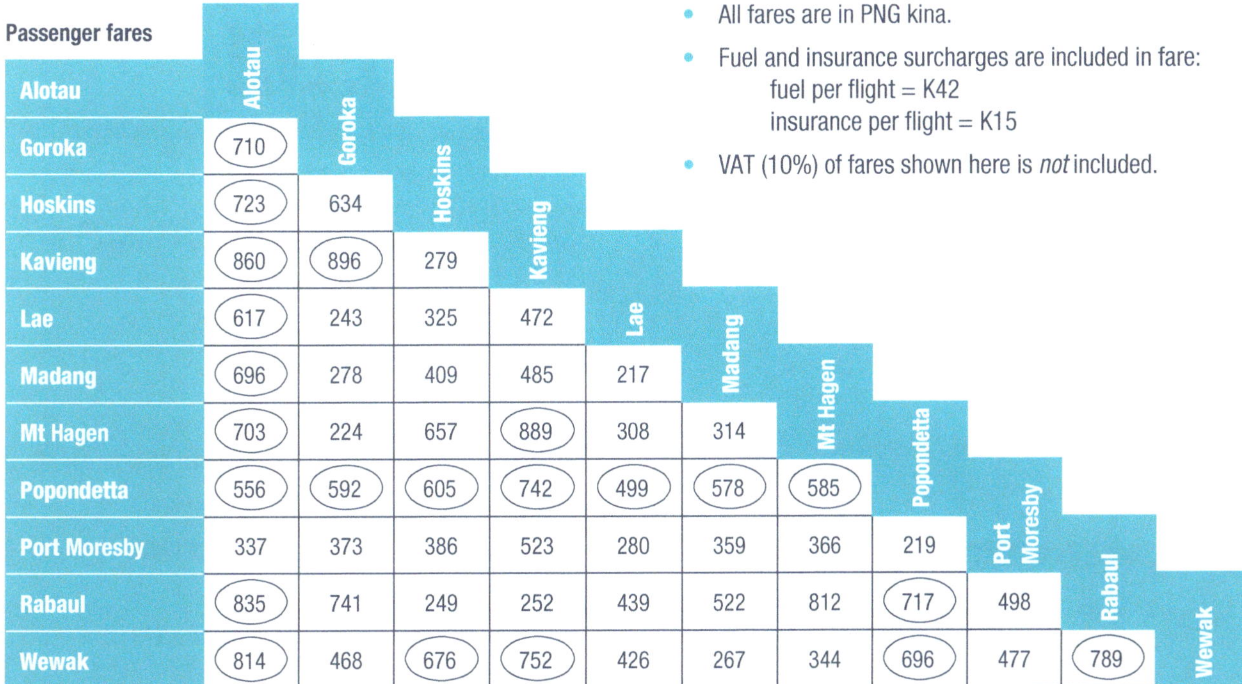

Passenger fares

	Alotau	Goroka	Hoskins	Kavieng	Lae	Madang	Mt Hagen	Popondetta	Port Moresby	Rabaul	Wewak
Alotau											
Goroka	710										
Hoskins	723	634									
Kavieng	860	896	279								
Lae	617	243	325	472							
Madang	696	278	409	485	217						
Mt Hagen	703	224	657	889	308	314					
Popondetta	556	592	605	742	499	578	585				
Port Moresby	337	373	386	523	280	359	366	219			
Rabaul	835	741	249	252	439	522	812	717	498		
Wewak	814	468	676	752	426	267	344	696	477	789	

- All fares are in PNG kina.
- Fuel and insurance surcharges are included in fare:
  fuel per flight = K42
  insurance per flight = K15
- VAT (10%) of fares shown here is *not* included.

All airfares include extra charges called surcharges. The airfares above include surcharges per flight for fuel (K42) and insurance (K15). This means that if a passenger books a return flight or their flight is not direct then each of these surcharges doubles. For example, the K710 for the journey from Alotau to Goroka includes a fuel surcharge of K84 (2 × K42) and an insurance surcharge of K30 (2 × K15) because it takes two flights to reach Goroka. The airfares above do **not** include Value Added Tax (VAT).

Use the table to help you answer these questions.

1 How much is the airfare (before adding VAT) from:

a Madang to Lae b Port Moresby to Kavieng c Rabaul to Alotau?

2 Which of the three airfares above is not a direct flight? What tells you this?

Help Box

To calculate the total cost of airfares we need to add VAT (a 10% tax imposed by the government) per flight. That is, for a one-way direct flight costing K279 we need to add 10% of K279.

There are two ways to calculate VAT:

1 Change 10% to a fraction and then multiply: 10% of K279 = $\frac{10}{100} \times \frac{279}{1}$ = K27.90

2 Change 10% to its decimal form and then multiply: 10% of K279 = 0.1 × 279 = K27.90

To work out the total fare, add the fare from the table to the calculated VAT.

For example, to travel from Kavieng to Hoskins: the fare from the table = K279
VAT = K27.90
So total fare = K306.90.

3 Use one of the methods described in the Help Box to calculate the total price for these fares.

a Port Moresby to Lae
b Mt. Hagen to Madang
c Rabaul to Goroka
d Wewak to Popondetta
e Wewak to Kavieng
f Port Moresby to Hoskins
g Mt. Hagen to Alotau
h Lae to Goroka
i Rabaul to Popondetta
j Wewak to Hoskins

Help Box

Value Added Tax (VAT) is a tax imposed on the sale of goods and services to raise revenue for the government. Customers pay 10% VAT on top of the sale price. The VAT from all sales is then paid to the government at regular periods by each shop or business.

Airlines can use the method below to work out the amount of VAT they have to pay to the government.

For example, the total fare from Kavieng to Hoskins is K306.90.

Divide K306.90 by 11 = K27.90.

K27.90 is the VAT to be paid to the government.

4 Calculate the amount of VAT that has to be paid on each of these direct one-way fares.

a K367.50 b K287.20 c K732.70
d K492.90 e K412.60 f K459.90
g K318.00 h K524.80

Some airlines offer discounts. Discounted tickets often have certain conditions such as travelling during specified times of the year, travelling on certain days and at certain times, paying for a ticket well in advance, and no changing of specified dates of travel. Discounts only apply to the base fare after all taxes and surcharges have been deducted.

Help Box

A discount is calculated in the same way as VAT.

For example, if a base fare is K195 and the discount is 20%, you can either:

1 change 20% to a fraction and then multiply:
20% of K195 = $\frac{20}{100} \times \frac{195}{1}$ = K39 or

2 change 20% to its decimal form and then multiply:
20% of K195 = 0.2 × 195 = K39

5 Calculate the discount on each of these base fares.

a 20% of K428 b 20% of K302
c 40% of K441 d 25% of K466
e 30% of K160 f 15% of K268
g 15% of K354 h 25% of K523
i 30% of K197

Help Box

It is useful to be able to estimate percentage discounts in your head.

For example, 20% of K428.

Step 1: In your head, calculate 10% of K428 = K42.80 and round it to K43.

Step 2: 20% is two times 10%, so in your head double K43 which is K86.

In question 5a above you found that the actual answer is K85.60.

6 Estimate these percentage discounts.

a 20% of K164 b 30% of K235
c 20% of K432 d 40% of K216
e 15% of K343 f 15% of K184
g 25% of K517 h 30% of K493
i 25% of K277

Use the information on the fares chart on page 75 to help you answer the following questions. Remember to include VAT in calculations where necessary.

7 What is the total cost for four adults to travel from Port Moresby on a direct flight to Hoskins?

8 A tourist flies from Port Moresby to Madang and then on to Rabaul.

a How much would the flight cost?

b Would it be cheaper to go to Rabaul first and then to Madang?

c What is the difference between the two fares?

9 A child's fare (2–11 years) from Mt. Hagen to Lae is 75% of the adult fare before VAT is added.

a If the adult fare before VAT is K308, what is the child's fare?

b What will each fare (adult and child) be after VAT is added?

10 Calculate the total cost for two adults and two children to fly from Madang to Rabaul. Children's fares are 75% of adult fares before VAT.

Lesson  5 Where will we stay?

Accommodation is a major travel expense for tourists. It can vary depending on the type and standard of accommodation required. In Papua New Guinea there are resorts, hotels, motels, guesthouses, lodges and village stays that provide accommodation for tourists.

Just like airfares, all transactions in hotels, restaurants, bars and shops are subject to 10% VAT, which is usually included in the prices quoted. Other ways of saying this are: VAT is applied in the ratio 1:10; VAT is equal to $\frac{1}{10}$ of the charge; for every 10 kina that is charged, VAT of 1 kina is added to the charge.

1 Which of these charges and VAT have been applied in the ratio 1:10?

a charge of K150 and VAT of K15

b charge of K200 and VAT of K20

c charge of K90 and VAT of K10

d charge of K110 and VAT of K11

e charge of K55 and VAT of K5

f charge of K175 and VAT of K17.50

2 How much VAT in the ratio 1:10 would be added to these rates?

a K240 b K120 c K95

d K146 e K224

Help Box

In the previous lesson you saw that the amount of VAT included in a price could be calculated by dividing by 11. This is because VAT is applied in the ratio 1:10, which means there are 11 parts in total: 1 part VAT and 10 parts cost.

For example, if the total cost to a tourist is K181.50, then VAT = K181.50 ÷ 11.

$$\frac{181.50}{11} = 16.50$$

So the amount of VAT included in the total is K16.50.

The cost before VAT is K165 (10 × K16.50).

The cost after VAT is K181.50 (11 × K16.50).

The ratio is K16.50:K165, and when simplified it is K1:K10 or 1:10.

3 This chart shows how much tourists paid for rooms at a hotel. Copy it into your workbook and complete it by working out the VAT and the cost before VAT is added.

	Cost of room before VAT	VAT	Total cost to tourist
a			K198
b			K236.50
c			K385
d			K209
e			K170.50
f			K159.50

4 What is the total amount of VAT that the owner of the above hotel has to pay to the government for the rooms shown in the chart?

5 Copy this chart into your workbook and complete the gaps.

	Cost of room before VAT	VAT	Total cost to tourist
a	K95		
b		K24	
c		K37.50	
d	K275		
e	K160		
f		K41.50	

Accommodation prices vary greatly depending on the type and standard. The chart below shows prices for different types of rooms at a top quality hotel in Papua New Guinea. VAT of 10% has to be added to the given prices.

	Standard Room	Premium Room	Executive Room
Single	K310	K564	K682
Twin (Double)	K355	K608	K727
Triple	K400	K652	K771

Answer these questions using the chart.

6 What is the total cost of the following rooms when VAT is added?

a a standard single room b an executive twin room

c a premium twin room d a standard triple room?

7 If two people shared the cost of these rooms, how much would each person pay after VAT was added?

a a standard twin room b a premium twin room

c an executive twin room

8 If three people shared the cost of these rooms, how much would each person pay after VAT was added?

a a standard triple room b a premium triple room

c an executive triple room

9 Draw a chart to show the new prices if the hotel gives a 20% discount before VAT is added for the month of May.

Some tourists prefer to stay in serviced apartments that are cleaned each day and that have their own kitchen and laundry facilities. Prices are shown below for a two-bedroom serviced apartment in Port Moresby.

Overnight	K420/night plus 10% VAT
2–6 nights	K380/night plus 10% VAT
7–31 nights	K320/night plus 10% VAT

10 Use the prices above to calculate the cost of each of the following stays.

a 1 night b 3 nights c 10 nights

d 5 nights e 15 nights f 21 nights

g 8 nights h 28 nights

11 In the serviced apartment:

a Is it cheaper to stay for 6 or 7 nights?

b What is the difference in price?

12 Four people stayed at the apartment for 4 nights and shared the cost.

a What is the total amount each person paid?

b What does this work out to per person per night?

13 Three people stayed at the apartment for 7 nights and shared the cost.

a What is the total amount each person paid?

b What does this work out to per person per night?

Lesson 6 What will we do?

There are hundreds of dive sites around Papua New Guinea and many tourists choose to stay on a dive boat. As well as providing accommodation and meals, all boats have experienced crews and offer up to five dives per day.

The prices below are per person per night for a boat that takes 12 passengers in seven cabins. All meals and dive tanks are included in the price but tourists need to add 10% VAT.

Cabin	Number of people in cabin	Price (per person) per night (K)	10% VAT (per person)	Total cost (per person)
1	2	K1040		
2	2	K1040		
3	2	K1040		
4	2	K965		
5	2	K965		
6	1	K1248		
7	1	K1248		

1 Copy the table into your workbook and complete the last two columns.

2 If every bed was taken:

- a How much VAT is payable to the government per night?
- b How much money would the dive operator take per night?

3 The cost of food has to come from the money taken by the dive operator. What other costs do you think have to be paid from this money?

4 If every bed is taken for five nights:

- a What is the total amount of money including VAT that the dive operator takes?
- b How much VAT is payable to the government?

5 A group of nine tourists hired cabins 1, 3, 4, 5 and 6.

- a How much did it cost the group in total per night?
- b If they shared the total cost equally, how much would each person pay per night?
- c How much of each person's share per night was VAT?

Many tourists come to Papua New Guinea specifically to walk the Kokoda Track. As many as 20 businesses are engaged in guiding trekkers along the track. Super fit people can cover the track in 6 days but 9 or 10 days is more practical.

The chart below shows the total cost per person for a 9-day trek by three different trekking companies. Each cost includes meals, the flight from Port Moresby to Kokoda, road transfers, fees and taxes (including VAT), an experienced guide, and group porters.

	Company A	Company B	Company C
Trek	K4630	K6370	K4990
Hire of personal porter (optional)	K1275	K1125	K1310

6 Which company is the most expensive for the trek and the porter combined?

7 If 15 people decided to trek the Kokoda Track with Company C and each person hired a personal porter, what would be the total cost for the group?

8 If 18 people chose to trek with Company B and two-thirds of them hired a personal porter:

a How much money would the company receive for the trek and porter hire?

b How much money in total would the company pay to personal porters?

9 A porter working for Company A did 14 treks during the year and carried an average of 14.5 kg each time. Each bag was a different weight.

a How much money did the porter earn from the treks?

b What might the exact weight have been of each bag carried by the porter?

The Sepik River is probably the most well known attraction of Papua New Guinea. Tourists can travel along it by boat, stopping at villages along the way. Travel is by way of dugout motor-canoes and motorboats, or the 'Sepik Spirit', a cruise boat with facilities for 18 passengers in nine 2-bed cabins.

Overseas tourists often organise cruises on the 'Sepik Spirit' before they arrive in Papua New Guinea, and different travel companies charge different prices. The prices shown in the table are those charged per adult by two overseas travel companies for a 4-day and 3-night cruise.

	Twin share	Single
Company X	K5170	K7680
Company Y	K3825	K5735

These prices include accommodation and meals on board, village sightseeing tours and government taxes. Children are charged 50% of the adult rate.

10 If a person chooses a single cabin, how much more will they pay than if they share a 2-bed cabin (twin share)? Calculate the extra for travel with each company.

11 Why do you think the cruise operators charge more for a person who books a cabin for themself than for people who share a cabin with another person?

12 A family of four, two adults and two children, book two 2-bed cabins with Company Y. How much will the cruise cost them?

13 There are nine cabins each accommodating two people. If all 18 beds are taken by adults, how much money will be taken by:

a Company X b Company Y?

14 Company X books four twin share cabins and three single cabins on the 'Sepik Spirit' for a group of eleven adults. What is the total cost for the group?

Challenge

Of the K3825 paid per adult for a twin share, how much do you think is spent on each of these: food, village tours, government taxes?

What other things have to be paid for?

How much profit do you think the owner of the cruise boat makes out of every K3825?

15 A group of five adults wants to spend 3 nights on a dive boat and then go on a 4-day and 3-night Sepik cruise. They do not mind sharing a cabin with other tourists.

a Use the information in this lesson to plan and calculate a total cost for the above experiences (excluding airfares between places).

b If the five adults shared the cost that you calculated, what would each person pay?

c What is the maximum amount the tourists would pay if only the most expensive options were available and the fifth person did not want to share?

Lesson 7 Patterns in handcrafts

Many modern handcrafts in Papua New Guinea have been produced for the tourist market and are sold at local markets or in hotel gift shops. Traditional crafts can still be bought in local villages and craft shops.

Shapes that have rotational symmetry are often used in handcraft design. You can decide if a shape has rotational symmetry by following the steps below.

Step 1: Make two copies of a shape and cut them out.

Step 2: Take one shape, stick it into your workbook and place the other shape directly on top of it. Mark the top of this shape with a cross.

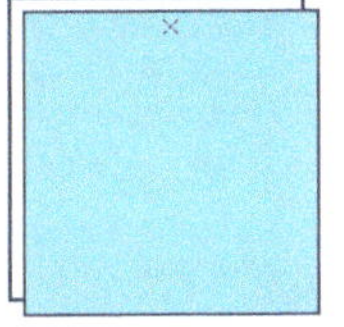

Step 3: Use a pencil point or similar to hold the shapes at the centre point and rotate the top shape until the cross is back where it started and the top shape lies exactly over the other again. This centre point is called the *point of rotation*.

Step 4: Count the number of times the top shape fitted onto the other shape during one complete turn.

Help Box

The square shown above fits onto a congruent square four times, so is said to have **rotational symmetry of order 4**. If a shape only fits onto a congruent shape when a whole turn has been completed then it does not have rotational symmetry.

1 Make two copies of each of the following shapes and follow the procedure above to decide which of them have rotational symmetry.

A

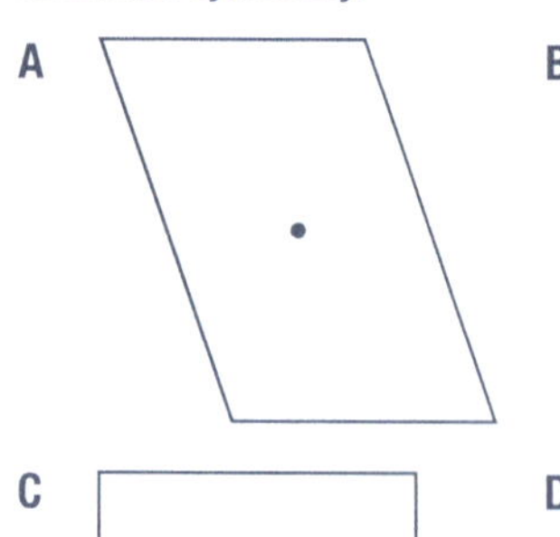

B

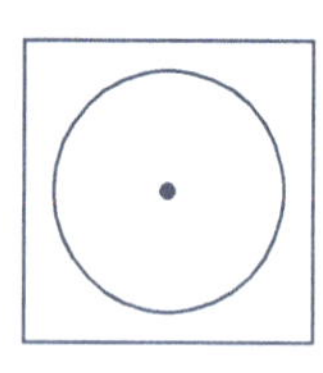

C

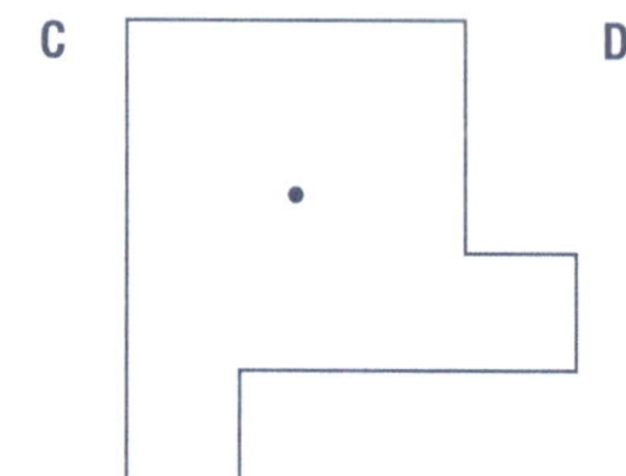

D

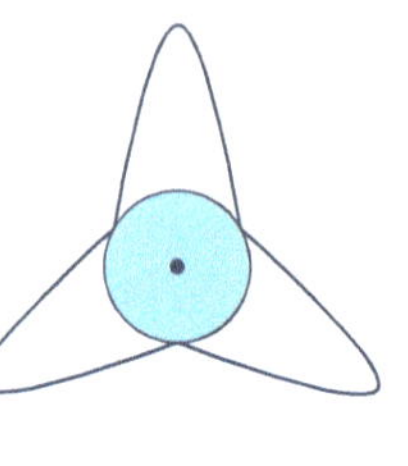

E

F

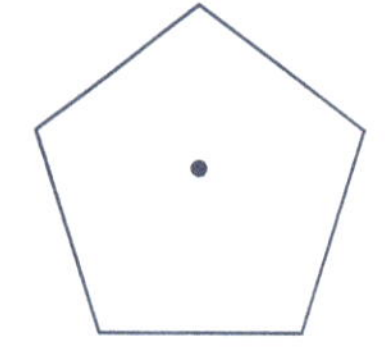

2 Use the shapes made in question 1 to complete the following:

a shape **A** has rotational symmetry of order _

b shape **B** has rotational symmetry of order _

c shape **C** has rotational symmetry of order _

d shape **D** has rotational symmetry of order _

e shape **E** has rotational symmetry of order _

f shape **F** has rotational symmetry of order _

3 Which of the above shapes does not have rotational symmetry?

4 Draw a shape that has rotational symmetry of order:

a 2 b 3 c 4 d 5

5 Copy some shapes from local handcrafts. Make two copies of each shape and rotate one shape on the other to see if the shapes have rotational symmetry. Make a chart to show your findings.

Many handcraft designs are made of shapes that tessellate. Some of these tessellated designs or patterns also have rotational symmetry, even though the individual shapes that make up the design do not have rotational symmetry.

6 Follow the steps below to make your own rotational tessellation.

Step 1: Rule and cut out a cardboard square at least 6 cm × 6 cm. Draw a shape on one side (Side AB).

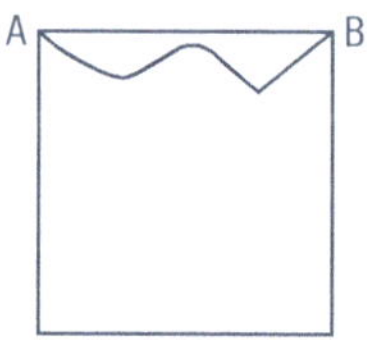

Step 2: Cut out the shape and rotate it 90° anticlockwise from Point B. Join it to the adjacent side of the square with tape. Place a small mark at Point B.

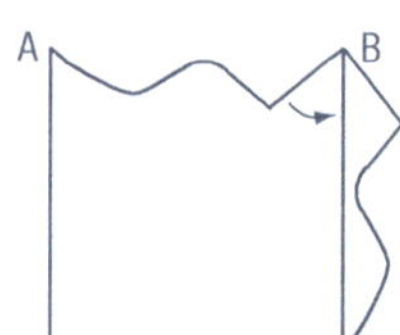

Step 3: Use the new shape as a template. Rotate the shape around Point B to make a tessellating design.

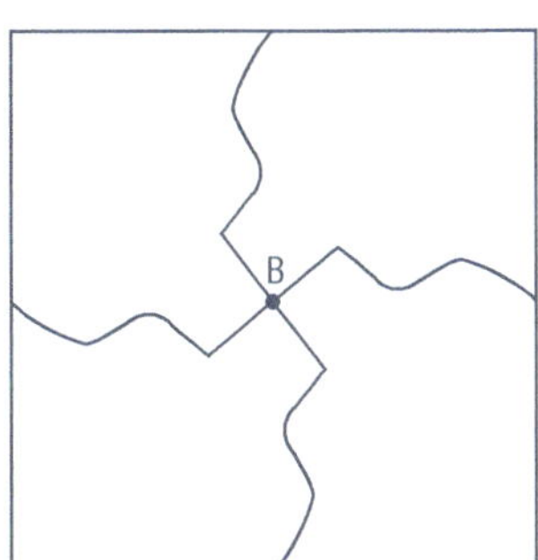

Help Box

To make a rotational tessellation, you must start with a square, an equilateral triangle or a regular hexagon.

When you have drawn a shape on one side and cut it out, join it to an adjacent side.

For example, here is a square, an equilateral triangle and a regular hexagon.

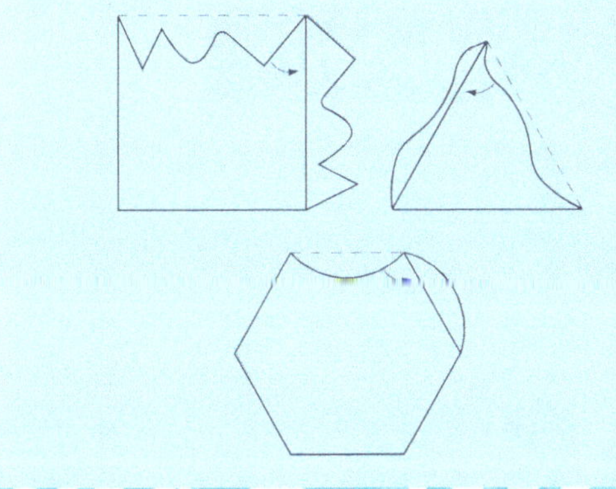

7 Use a ruler and protractor to construct the following shapes:

- a a square
- b an equilateral triangle
- c a regular hexagon

8 Create a new shape (different from those in the Help Box) from each of the constructed shapes.

9 Use the new shapes as templates to make three tessellating designs.

10 Identify a point of rotation on each of your designs. What is the order of rotational symmetry for each tessellation?

Help Box

To make even more interesting tessellations, you can cut out shapes on two sides of a square and two or three sides of a regular hexagon. Note each shape must be joined to an adjacent side. Here are some examples.

11 Construct either a square or a hexagon.

- a Create a tessellating shape by cutting and joining shapes on two sides of the square or at least two sides of the hexagon.
- b Use your new shape to create a tessellating design. Decorate it how you want.

12 Find some tessellating designs on local handcrafts. Which ones have rotational symmetry? How do you know?

Lesson 8 Renting a car

Rental cars are available in Papua New Guinea even though car hire is expensive and there is a limited road network. Cars are supplied to renters with a full fuel tank and must be returned with a full tank.

The following table shows the approximate fuel consumption and cost per day for five different types of vehicles.

Vehicle	Fuel consumption (L per 100 km)	Cost per day
4-cylinder manual	6.5	K198
4-cylinder automatic	7.1	K216
6-cylinder manual	8.9	K255
6-cylinder automatic	9.6	K284
6-cylinder automatic wagon	10.2	K299

Help Box

We can see that a 4-cylinder manual car uses about 6.5 litres for every 100 kilometres travelled. To calculate the number of litres it uses for any distance we can use the formula: $\frac{\text{fuel consumption}}{100} \times \text{distance travelled}$

For example, for a distance of 250 km:

$$\frac{6.5}{100} \times 250 = 16.25 \text{ L}$$

So, if we drive a 4-cylinder manual car for 250 kilometres we will use 16.25 litres of petrol. (This is an approximate amount as many things including speed, tyre pressure and temperature affect fuel consumption.)

1 How much fuel would be used to travel these distances?

- a 375 km in a 6-cylinder automatic
- b 160 km in a 4-cylinder manual
- c 285 km in a 4-cylinder automatic
- d 485 km in an automatic wagon
- e 320 km in a 6-cylinder manual
- f 215 km in a 4-cylinder automatic
- g 154 km in an automatic wagon
- h 389 km in a 6-cylinder automatic

2 If fuel costs K2.77 per litre, what would the fuel cost be for each of the distances in question 1?

3 Find the current cost of fuel in your area and use it to calculate the fuel cost for each of the distances in question 1.

Help Box

We can estimate the amount of fuel needed for any distance if we know how many litres of fuel are used per 100 kilometres.

For example, to travel 310 km in a 4-cylinder automatic. We know that a 4-cylinder automatic uses 7.1 L for every 100 km travelled, so if we round 7.1 L to 7 and 310 km to 300, we can estimate by calculating 7 × 3 (3 lots of 100 km) = 21 L.

4 Use the method above to estimate the amount of fuel needed for these distances.

- a 190 km in a 6-cylinder wagon
- b 213 km in a 4-cylinder automatic
- c 487 km in a 6-cylinder manual
- d 327 km in a 6-cylinder automatic
- e 508 km in a 4-cylinder manual
- f 479 km in a 4-cylinder automatic

5 Estimate the amount of fuel that would be used in the following situations:

a a distance of 295 km in a car with a fuel consumption of 5.1 L per 100 km

b a distance of 507 km in a car with a fuel consumption of 8.7 L per 100 km

c a distance of 379 km in a car with a fuel consumption of 11.3 L per 100 km

d a distance of 196 km in a car with a fuel consumption of 7.8 L per 100 km

e a distance of 423 km in a car with a fuel consumption of 6.3 L per 100 km

f a distance of 214 km in a car with a fuel consumption of 7.2 L per 100 km

Help Box

We can use the previous information to calculate the fuel consumption per 100 km of any car if we know the distance travelled and the amount of fuel that was used.

To calculate the fuel consumption per 100 km of a car that travels 263 km and uses 30 L of fuel, we use the formula:

$$\text{fuel consumption per 100 km} = \frac{\text{number of litres}}{\text{distance}} \times 100$$

For example, $\frac{30\text{ L}}{263} \times 100 = 0.114 \times 100$
$= 11.4\text{ L per 100 km}$

6 Calculate the fuel consumption per 100 km of a car that travels:

a 415 km and uses 37 L of fuel

b 187 km and uses 15 L of fuel

c 238 km and uses 21 L of fuel

d 366 km and uses 25 L of fuel

e 152 km and uses 9 L of fuel

f 279 km and uses 19 L of fuel

Challenge

Calculate the distance travelled by a car with a fuel consumption of 9.2 L per 100 km that uses 30 L of fuel.

Refer to the table on page 84 to help you answer these questions.

7 A tourist rented a 4-cylinder automatic car for 3 days. Before returning the car he filled it with 30 litres of petrol at K2.83 per litre. What was the total cost for rental and petrol?

8 A group of five tourists hired a 6-cylinder wagon for 4 days. They spent K137.50 on petrol during the trip and put another 35 litres, at K2.92 per litre, in the tank before returning the wagon.

a What was the total cost for rental and petrol?

b If the tourists shared the cost equally, how much did each person pay?

9 In one week a car rental company rented out four 4-cylinder manual cars for 3 days each, two 6-cylinder automatics for 4 days each, three 4-cylinder automatics for 2 days each and a 6-cylinder wagon for 6 days.

a How much money did the rental company take for the week?

b Lict como of the costs an owner of a car rental company might have.

Lesson 9 Planning a journey

A travel agent wrote this eight-day itinerary for a group of tourists.

8-day itinerary

Day 1:	arrive in Port Moresby sightseeing around Port Moresby
Day 2:	plane from Port Moresby to Rabaul sightseeing in Rabaul
Day 3:	tour of volcanoes in Rabaul
Day 4:	plane from Rabaul to Madang sightseeing in Madang
Day 5:	at leisure in Madang
Day 6:	bus trip from Madang to Goroka overnight stay in Goroka
Day 7:	return by bus from Goroka to Madang
Day 8:	plane from Madang to Port Moresby to catch international flight

1 Answer these questions about the itinerary.

a How many places will the tourists visit?

b How many nights will they have in Madang?

c How many flights will they have once they arrive in Papua New Guinea?

d How many nights' accommodation in Papua New Guinea will the travel agent need to book for the tourists?

2 Imagine that you are planning a 10-day trip in Papua New Guinea for a group of tourists. The tourists are interested in sightseeing and swimming but do not want to go trekking, fishing or diving.

a Write an itinerary for a suggested 10-day trip. Include the method of travel between each place. Assume that the tourists arrive in Port Moresby on Day 1 and depart from Port Moresby on Day 10.

b Write an alternative itinerary that would be suitable for the same group of tourists. It can include some of the same places that are in your first itinerary.

3 Sketch a map of Papua New Guinea.

a Label all the places the tourists will visit according to your two itineraries.

b Join the places in the order that the tourists will visit them according to your first itinerary.

c Use a different colour pen to join the places in the order that each will be visited according to your alternative itinerary.

Some tourists were visiting Esther's town, so Esther drew this scale map of her town and marked the sights that she thought the tourists would be interested in. She drew a line to show a walking tour that the tourists could take to see the sights.

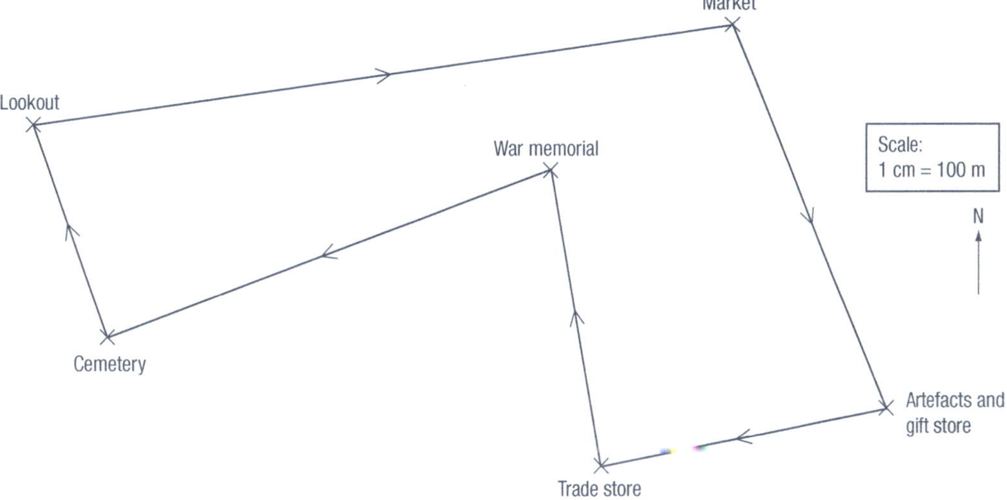

4 Answer these questions about the map.

- **a** How far is it from the lookout to the market?
- **b** How do tourists know which way to walk?
- **c** What direction will the tourists travel from the war memorial to the cemetery?
- **d** About how far will the tourists walk if they do the whole tour?
- **e** If tourists walk at an average speed of 3.5 kilometres per hour and spend an average of 30 minutes at each place, about how long will the walking tour take them?

5 Draw a scale map of all or part of your town or village and mark at least five sights that you think tourists might be interested in seeing. Don't forget to include the scale, a direction arrow and arrows to show which way to walk.

6 Answer these questions about your map.

- **a** How far is your suggested walking tour?
- **b** How long will it take tourists to complete your tour if they walk at 3.5 kilometres per hour and spend 30 minutes at each sight?

Lesson 10 Holiday or work?

The table below shows the total number of visitors to Papua New Guinea from 1996 to 2005. The table also shows how many of the total number of visitors were here for a holiday and how many were here on business.

Year	Total	Holiday	Business
1996	61 392	15 217	18 682
1997	65 960	16 965	19 429
1998	67 545	22 616	15 986
1999	67 816	19 370	17 175
2000	58 448	13 792	16 441
2001	54 235	13 896	17 331
2002	53 762	15 279	15 492
2003	56 282	15 112	16 427
2004	60 715	17 280	18 663
2005	68 450	17 584	22 947

Source: National Statistical Office of Papua New Guinea

1 Answer these questions about the table.

a In which year did Papua New Guinea have the most number of visitors?

b In which year did Papua New Guinea have the least number of visitors?

c How many more visitors came in 1999 than 2000?

d How many more visitors came for business in 2005 than 2004?

e How many fewer visitors came for a holiday in 2002 than 2005?

f In 2005, how many visitors came for reasons other than a holiday or business?

2 List other reasons that people might have for visiting Papua New Guinea.

Help Box

In 2005, 17 584 of the total number of visitors to Papua New Guinea were on holiday. This figure can be written as a percentage of the total by following these steps.

Step 1: Write the number of visitors on holiday as a fraction of the total number of visitors: $\frac{17\,584}{68\,450}$

Step 2: Multiply this fraction by 100:

$$\frac{17\,584}{68\,450} \times \frac{100}{1} = \frac{1\,758\,400}{68\,450} = 25.68\%$$

Step 3: Round 25.68 off to one decimal place = 25.7%

25.7% of the total visitors to Papua New Guinea in 2005 were here on holiday.

3 Use the method shown in the Help Box to find the percentage of the total number of visitors to Papua New Guinea that were:

a on business in 2000

b on holiday in 1997

c on business in 2004

d on holiday in 2003

e on business in 2001

f on holiday in 1998

4 Which year had the greater percentage of the total number of visitors on holiday?

a 1996 or 1997

b 1999 or 2000

c 2003 or 2004

5 Which year had the greater percentage of the total number of visitors on business?

a 1997 or 1998

b 2001 or 2002

c 2004 or 2005

6 In which years was the percentage of holiday visitors less than a quarter of all visitors?

In 2004, 60 715 visitors arrived in Papua New Guinea. In 2005, this number increased to 68 450. This means that 7735 more people arrived in 2005 than in 2004. Follow the instructions in the Help Box below to work out this increase in the number of visitor arrivals as a percentage increase.

Help Box

To calculate an increase in visitor arrivals as a percentage increase, follow these steps.

Step 1: Find the difference between the number of arrivals in 2004 and 2005:
68 450 – 60 715 = 7735

Step 2: Write the difference as a fraction of the total number of visitors in 2004: $\frac{7735}{60715}$

Step 3: Multiply this fraction by 100:
$\frac{7735}{60715} \times \frac{100}{1} = \frac{773500}{60715} = 12.74\%$

Step 4: Round 12.74 off to one decimal place = 12.7%

Visitor arrivals increased by 12.7% from 2004 to 2005.

7 Calculate the percentage increase in total visitor arrivals from:

a 2003 to 2004 b 1996 to 1999

c 2002 to 2005

8 Calculate the percentage increase in the number of holiday arrivals from:

a 2002 to 2005 b 1997 to 1998

c 1996 to 2005

Help Box

Where the number of arrivals decreases, we can calculate the percentage decrease in a similar way. For example, in 1998 there were 22 616 holiday arrivals and in 2005 this had decreased to 17 584.

Step 1: Find the difference between the number of holiday arrivals in 1998 and 2005:
22 616 – 17 584 = 5032

Step 2: Write this as a fraction of the total number of holiday arrivals in 1998 and multiply it by 100:
$\frac{5032}{22616} \times \frac{100}{1} = \frac{503200}{22616} = 22.25\%$

Step 3: Round 22.25 off to one decimal place = 22.3%

Holiday arrivals decreased by 22.3% from 1998 to 2005. This can also be written as –22.3%.

9 Calculate the percentage decrease in total visitor arrivals from:

a 2000 to 2001 b 1999 to 2000

c 1998 to 2002

10 Calculate the percentage decrease in the number of business arrivals from:

a 1997 to 1998 b 1999 to 2002

c 2001 to 2003

Challenge

Calculate the percentage increase or decrease in the total number of visitor arrivals each year from 1996 to 2005 and draw a graph to display your findings.

Learning Unit 2 Travelling Overseas

Strand: Number and Application

Decimals	Outcome 8.1.2	Use decimals to solve real life problems
Decimals and Percentages	Outcome 8.1.4	Solve problems in any situation that involves percentages
Ratios and Rates	Outcome 8.1.6	Apply rates to solve simple problems from real life

Strand: Space and Shape

Direction	Outcome 8.2.14	Use and read maps accurately
Maps and Coordinates	Outcome 8.2.16	Use longitude and latitude to locate places on a map

Strand: Measurement

Weight	Outcome 8.3.1	Solve real life weight problems with confidence and competence
Temperature	Outcome 8.3.3	Display temperatures including those from specialist thermometers
Time	Outcome 8.3.4	Recognise relationships between location and time
Time	Outcome 8.3.5	Use time–rate calculations

Strand: Chance and Data

Statistics	Outcome 8.4.1	Interpret information presented statistically
Accuracy and Error	Outcome 8.4.5	Represent levels of accuracy
Estimation	Outcome 8.4.7	Estimate results of calculations

Lesson 1: Introduction	Planning and organising an overseas trip
Lesson 2: Leaving Papua New Guinea	Introducing map projections Understanding and using the symbol for approximation (≈) Using scale to calculate distances on a map
Lesson 3: Where in the world is ...?	Understanding latitude and longitude Locating places given their latitude and longitude Using degrees and minutes to state latitude and longitude
Lesson 4: Temperature variations	Relating temperature to distance from the equator Interpreting temperature graphs and tables
Lesson 5: Time zones	Interpreting a world time zones map Determining time in other world cities based on local Papua New Guinea time
Lesson 6: Plane specifications	Using whole numbers and decimals to compare features of different aircraft Calculating flying time, distance and fuel consumption
Lesson 7: Pack your bag	Solving problems involving allowable baggage weights for airline passengers Calculating excess baggage rates
Lesson 8: What is my money worth?	Investigating exchange rates Converting between kina and other currencies
Lesson 9: Airfare deals	Solving money problems involving decimals Calculating percentage discounts Finding an amount of money when given a percentage of it

Lesson 1 Introduction

In this unit you will look at the mathematics behind some of the issues that are involved when travelling overseas to other countries. There are many things that have to be organised even before you can depart Papua New Guinea, such as applying for a passport. Different countries have different entry requirements so you need to be aware of these and plan in advance. Immunisations and travel insurance are other aspects that need consideration.

1 One of the major reasons for travelling overseas is for holidays. Spend some time thinking of other reasons people might want or need to travel overseas and make a list of them.

2 Look at a map of the world and choose three countries you would like to visit. List each country in order of preference and give your reasons for each choice.

3 Name some countries that are close to Papua New Guinea.

4 Name some countries that are a long way away from Papua New Guinea.

Anyone who wants to travel overseas must have a passport. This is an official travel document that identifies who you are and what country you come from. It is usually a small booklet that contains a description and photograph of the bearer.

5 Menari has decided that he wants to travel to Australia for a holiday. Find Australia in your atlas and name at least five cities that Menari could visit while there.

6 Menari has started to make a list of things he has to organise before he travels. Discuss this list with a friend. Copy it into your workbook and write other things that Menari should organise before going to Australia.

Things to do
- Apply for and obtain a passport
- Work out a departure and return date
- Decide which airline company he will travel with
- Purchase a plane ticket
- Check if immunisations are required
- Find out if any visas are required
- Organise travel insurance

While Menari is in Australia he has to decide to do three of the following activities:

- visit Uluru

- climb the Sydney Harbour Bridge

- visit the Great Barrier Reef

- visit the snow

- ride a tram in Melbourne

- see a koala and a kangaroo

6 Which three of these activities would you choose to do? Why?

Lesson 2 Leaving Papua New Guinea

As the world is shaped like a sphere it is difficult to accurately represent it, or parts of it, on a flat surface. A method called map projection is used to represent the curved surface of the Earth or part of the Earth on a flat surface. A drawback of this method is that some distortions of distance, direction, scale and size always occur. On many flat maps the scale is only constant between points along the equator as this is the only part of the curved surface that is not distorted when the map is flattened.

This map shows part of the world with some of the major cities marked.

Use the map on the previous page to help you answer the following questions.

1 There are three horizontal lines marked on the map.

a Write the name of each of the three lines.

b Which of the three lines is furthest north?

c Which of the three lines runs through Australia?

2 Name three cities that are south of the Tropic of Capricorn.

3 Name three cities that are between the equator and the Tropic of Cancer.

4 Use direction and the three marked lines to describe the position of these cities:

a Port Moresby, Jakarta and Nairobi

b Delhi, Shanghai and Tokyo

Help Box

You will notice that the scale marked near the map uses the symbol ≈. This is the symbol for approximation and means that 1 cm on the map is approximately equal to 1140 km on the ground.

It is impossible to measure distances accurately on this map because of the distortion produced by map projection and because the scale on the map is so small.

You will need a ruler and the map on the previous page to answer the following questions. Use the symbol for approximation in your answers.

5 What is the approximate distance between Port Moresby and the following places?

a Brisbane b Shanghai c Nairobi d Paris e Cairo

6 What is the approximate distance between these places?

a Johannesburg and Jakarta

b Baghdad and Bangkok

c London and Moscow

d Lagos and Istanbul

e Melbourne and Karachi

f Madrid and Brisbane

g Stockholm and Tokyo

h Rome and Johannesburg

7 Use the map to decide if these statements are true or false.

a The distance between Lagos and Istanbul is ≈ 4000 km.

b The distance between Baghdad and Manila is ≈ 3000 km.

c The distance between Copenhagen and Nairobi is ≈ 6300 km.

d The distance between Port Moresby and Athens is ≈ 11 400 km.

e The distance between Johannesburg and Calcutta is ≈ 4500 km.

f The distance between Moscow and Tokyo is ≈ 8500 km.

8 Arnold did a return trip between Port Moresby and Nairobi. Approximately how many kilometres did he travel?

9 Anna is going to fly from Port Moresby to Hong Kong and then on to London. Approximately how many kilometres will she travel?

10 Use the map and a ruler to plan a trip that is between 25 000 km and 30 000 km. You must start and finish at Port Moresby and can visit as many or as few places as you like.

Lesson 3 Where in the world is ...?

Lines of latitude and longitude are a grid system drawn on maps so that you can find places with accuracy. The lines of latitude are imaginary lines running in an east–west direction around the Earth and, as you saw in the previous lesson, include the equator, the Tropic of Cancer and the Tropic of Capricorn.

The lines of longitude are imaginary lines running in a north–south direction from pole to pole. Latitude and longitude are measured in degrees. The equator is 0 degrees and all other lines of latitude are north or south of the equator. The prime meridian is the line of longitude that is 0 degrees and all other lines of longitude are degrees east or west of the prime meridian.

1 Look at the map below that shows part of the world.

a Name three countries that the equator passes through.

b Name three countries that the prime meridian passes through.

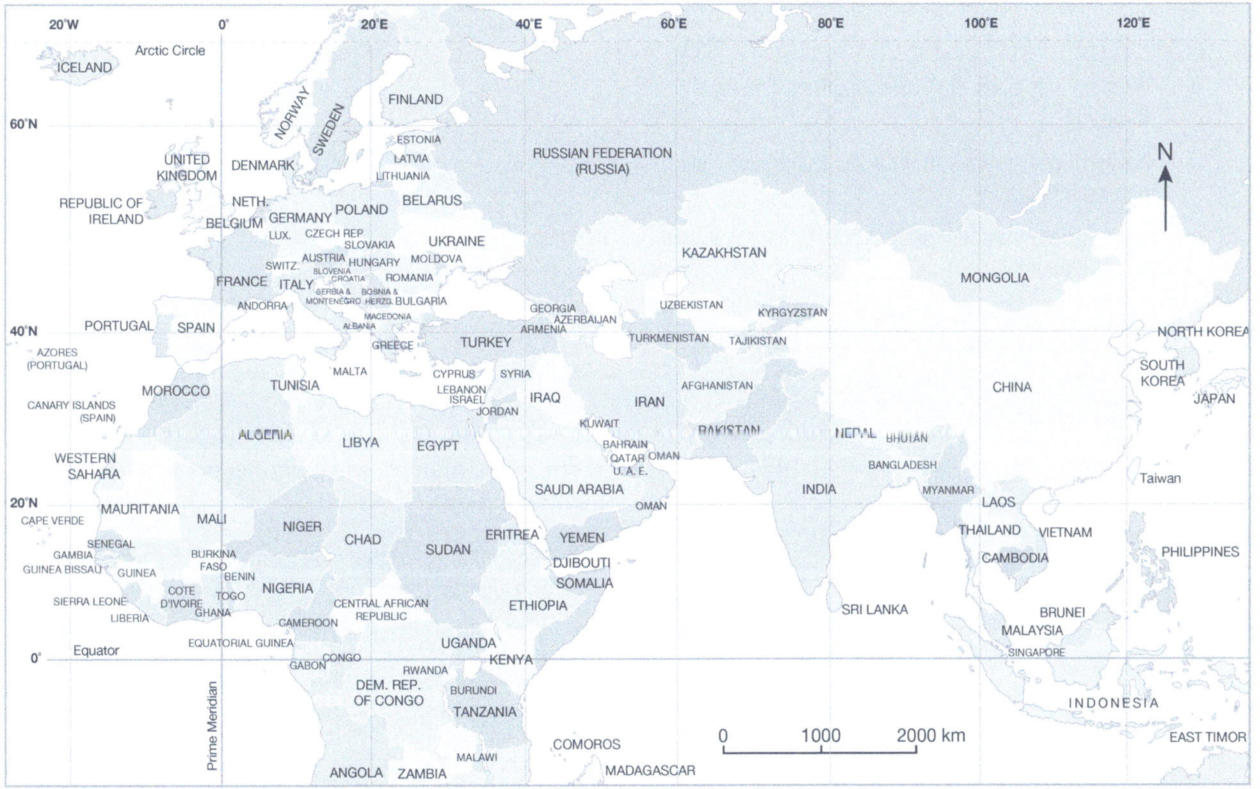

2 On the map on page 95 you can see lines of latitude north of the equator. There are a maximum of 90° of latitude to the north or the south of the equator.

a Name three countries that the 20°N latitude line passes through.

b Name three countries that the 60°N latitude line passes through.

3 Not every line of latitude is marked on the map. Use the lines that are marked to help you find one country that each of the following unmarked lines of latitude would pass through:

a 10°N b 30°N c 65°N

d 5°S e 50°N

4 The map also shows lines of longitude east and west of the prime meridian. There are a maximum of 180° of longitude to the east or the west of the prime meridian.

a What country does the 20°W longitude line pass through?

b Name three countries that the 40°E longitude line passes through.

c Name three countries that the 100°E longitude line passes through.

5 Not every line of longitude is marked on the map. Use the lines that are marked to help you find one country that each of the following unmarked lines of longitude would pass through:

a 30°E b 70°E c 10°W

d 45°E e 105°E

Help Box

Lines of latitude and longitude can be used together to specify an exact location. Each degree of latitude or longitude can be divided into 60 minutes. Minutes are shown by the symbol ′. For example, a longitude of 65°30′ E is exactly halfway between 65°E and 66°E.

When specifying an exact location, latitude is always given before longitude, for example, 35°N, 75°W. It is important to state the direction either north or south for latitude and either east or west for longitude.

The map below shows part of Vanuatu, a neighbouring country to the south-east of Papua New Guinea. The map is on a large scale and shows the lines of latitude and longitude in single degrees. The 30-minute marks (30′) have been placed to show you where the minutes fit with the degrees.

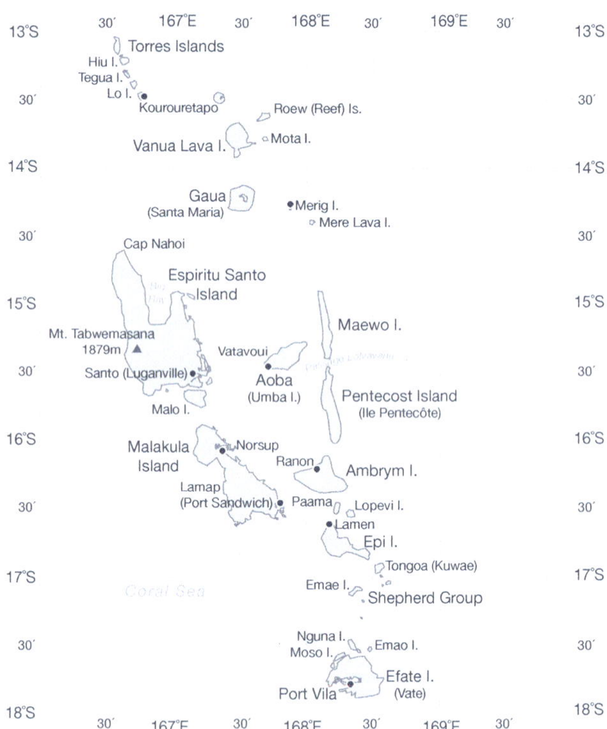

Use the 30-minute marks on the map to help you estimate the minutes when finding places or when giving locations.

6 Locate the place that is at each of the following latitudes and longitudes:

a 15°28′S, 167°45′E

b 17°45′S, 168°20′E

c 16°10′S, 168°5′E

d 14°15′S, 167°50′E

7 Give the location of the following places using both degrees and minutes:

a Norsup
b Lamen
c Lamap (Port Sandwich)
d Mt. Tabwemasana
e Santo (Luganville)
f Kourouretapo

This map shows the main islands of Fiji, a country to the east of Vanuatu. You will notice that the 180° line of longitude passes through Fiji. As there are a maximum of 180° to the east of the prime meridian, the next line of longitude is measured from the west of the prime meridian (179°W). The 180° line of longitude is both east and west and is written as 180° without a direction after it.

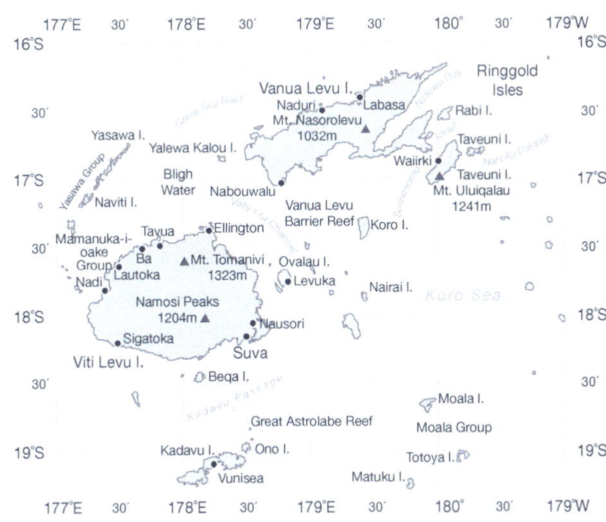

8 Locate the place that is at each of the following latitudes and longitudes:

a 18°10′S, 178°30′E
b 17°50′S, 177°25′E
c 16°50′S, 180°
d 17°S, 178°45′E

9 Give the location of the following places using both degrees and minutes.

a Levuka
b Ellington
c Vunisea
d Sigatoka
e Lautoka
f Labasa

Challenge

Use an atlas to help you find other cities in the world that are about the same latitude as Port Moresby.

Use an atlas to help you find other cities in the world that are about the same longitude as Port Moresby.

Lesson 4 Temperature variations

The temperature and temperature variation in a place largely depend on its position in relation to the equator. Places nearer the equator are generally warmer than places nearer the poles. At the equator the sun is high overhead all year so there is little variation in average temperatures throughout the year. Because Papua New Guinea is close to the equator, its temperature has little variation through the year. Other places that are close to the equator experience a similar small variation in temperature throughout the year.

1 Find a world map in an atlas.

a Name at least two countries or cities where you would expect the temperature to have little variation throughout the year.

b Write a sentence to explain why these countries or cities would have little temperature variation.

c Name at least two countries or cities where you would expect the temperature to vary quite a bit throughout the year.

d Write a sentence to explain why the temperature would vary in these places.

2 Look at the graphs below. They show the annual 24-hour average temperature in degrees Celsius for different cities.

a Which of the cities would you expect to be close to the equator? What tells you this?

b Which of the cities would you expect to be a long way away from the equator? What tells you this?

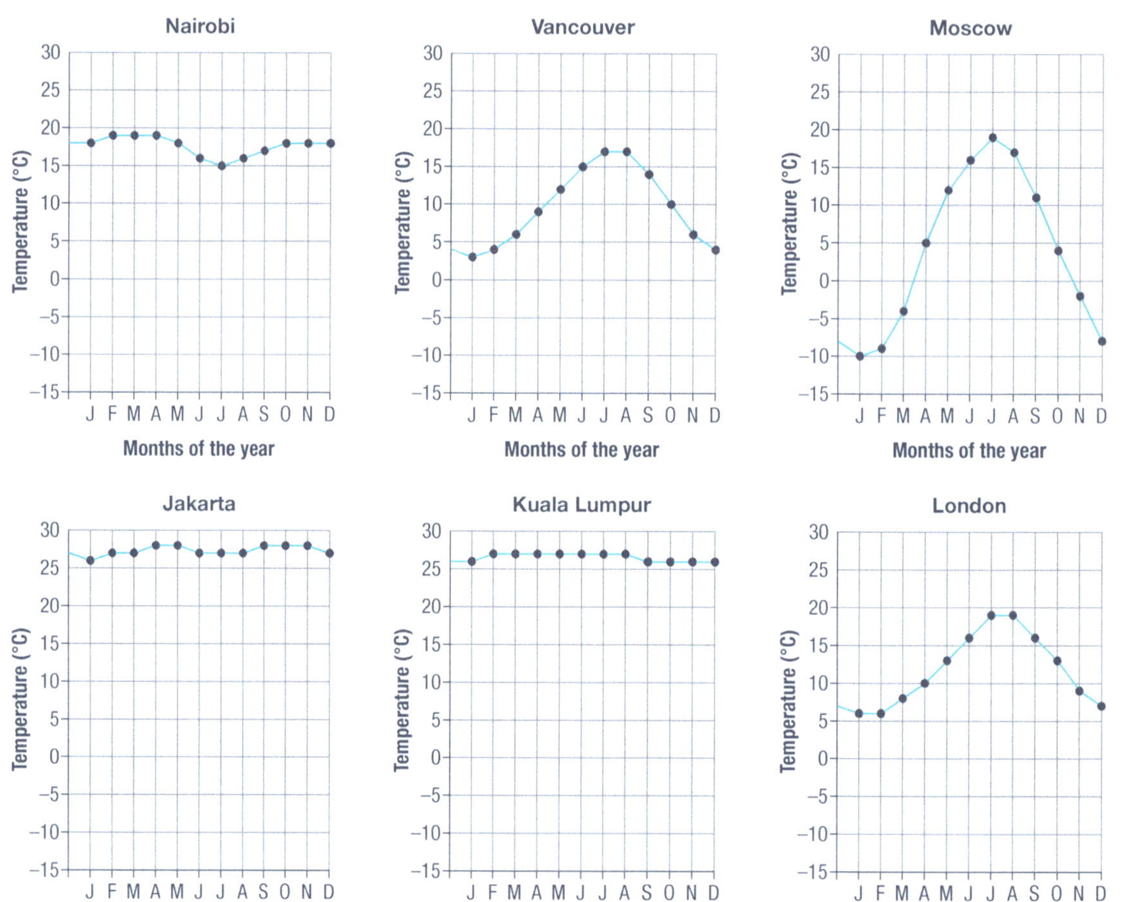

Temperature records are kept for many years and then averaged, so the information in the graphs opposite provides us with a reasonably accurate picture of expected mean temperatures in various places at various times of the year. However, the graphs do not give us data on the maximum or minimum temperatures for each month. For example, the mean temperature for Nairobi in January is 18°C but the extremes of temperature each day could be much higher or much lower. When travelling we need to know the daily extremes of temperature so we know what clothing to take with us.

3 The following table gives the mean maximum and mean minimum temperatures for the city of Brisbane in Australia.

	Jan	Feb	Mar	Apr	May	June	July	Aug	Sept	Oct	Nov	Dec
Mean maximum (0°C)	29	29	28	26	24	21	21	22	24	26	27	29
Mean minimum (0°C)	21	21	20	17	14	11	9	10	13	16	18	20

a Which months have the greatest variation in temperature?

b Which months have the least variation in temperature?

c If you were visiting Brisbane in July, what type of clothes would you pack?

d If you were visiting Brisbane in January, what type of clothes would you pack?

4 The following table gives the mean maximum and mean minimum temperatures for the city of Paris in France.

	Jan	Feb	Mar	Apr	May	June	July	Aug	Sept	Oct	Nov	Dec
Mean maximum (0°C)	7	8	12	15	19	22	24	25	21	16	10	8
Mean minimum (0°C)	3	3	5	7	11	13	16	15	13	9	5	4

a Which month has the greatest variation in temperature?

b Which months have the least variation in temperature?

c In which months would you need to pack warm clothes if you were visiting Paris?

5 Use the data from the two tables above to construct two line graphs, one for Brisbane and one for Paris. Each graph will contain two sets of data: the mean maximum monthly temperatures and the mean minimum monthly temperatures. Make sure you use the same scale on each graph so that you can compare them.

6 Look at the two graphs you have constructed.

a Describe the difference between the curves of the lines in each graph.

b Which city has its coolest temperatures in June, July and August?

c Describe the temperature of Paris in December, January and February.

d In which month would you prefer to visit Paris? Why?

e In which month would you prefer to visit Brisbane? Why?

7 Find a map of the world in an atlas and locate Paris and Brisbane.

a Use degrees to record the approximate latitude for Paris, including the direction.

b Use degrees to record the approximate latitude for Brisbane, including the direction.

The equator divides the Earth into the northern and southern hemispheres. Paris is in the northern hemisphere and Brisbane is in the southern hemisphere. Because of the Earth's movement around the sun, places in the northern hemisphere experience winter at the same time as places in the southern hemisphere experience summer. As we have seen, places nearer to the equator do not experience a large temperature variation because the sun is high overhead all year round.

8 Look at the following temperature graphs.

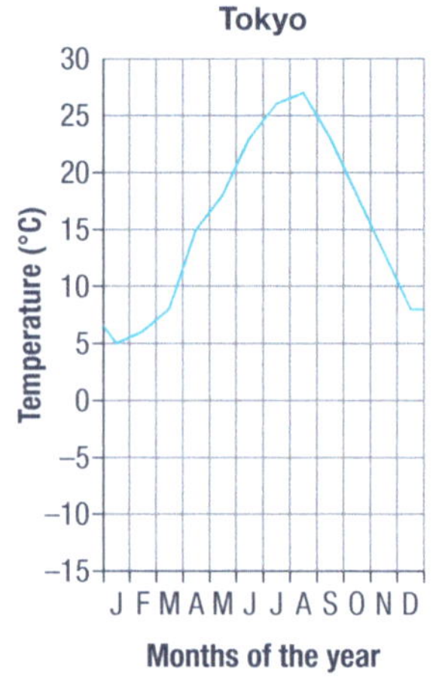

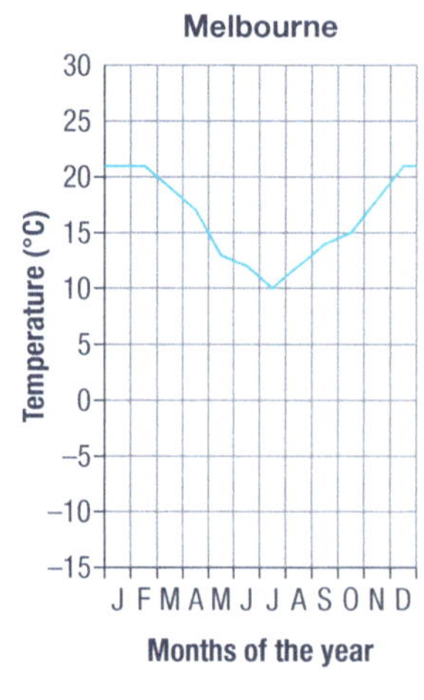

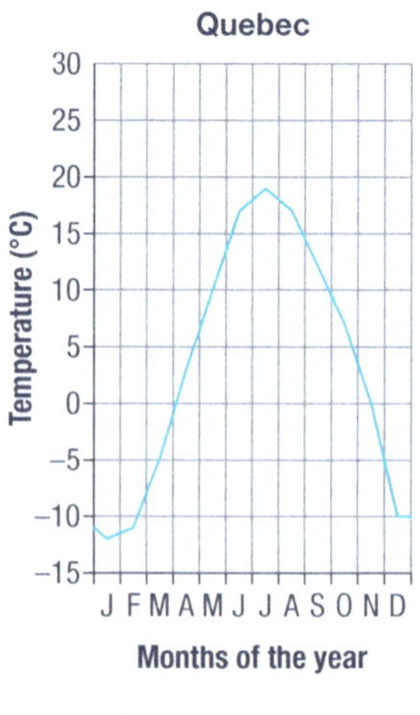

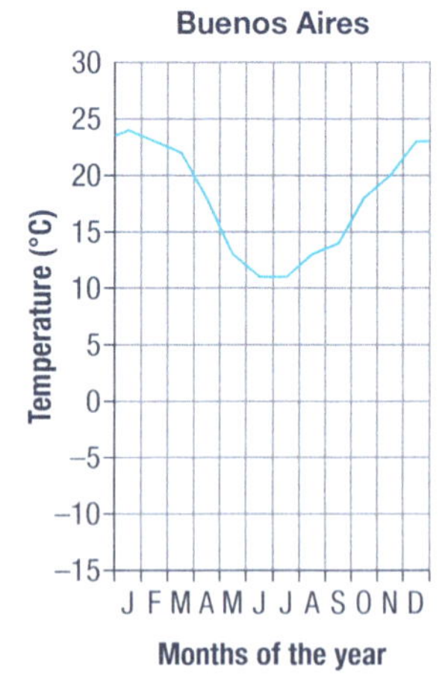

a Which of the four cities are in the northern hemisphere? What tells you this?

b Which of the four cities are in the southern hemisphere? What tells you this?

c Which city experiences the coldest winters?

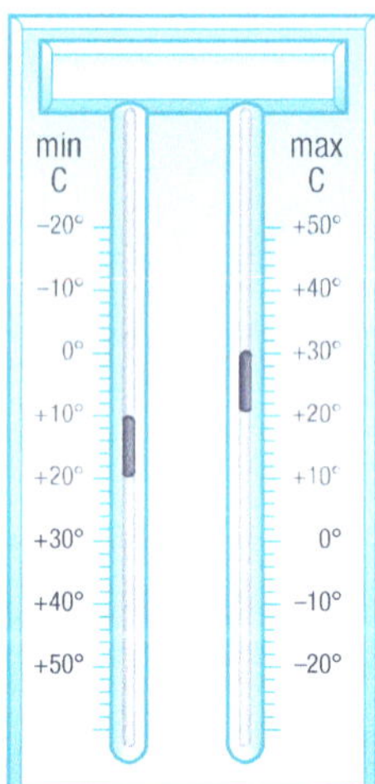

9 Ask your teacher for a maximum and minimum thermometer. Set it up in your school and use it to measure the day's maximum and minimum temperature.

a Record the maximum and minimum temperatures from the thermometer each day for one month.

b At the end of the month calculate the average maximum temperature and the average minimum temperature for that month.

c Calculate the range of maximum temperatures for the month.

d Calculate the range of minimum temperatures for the month.

10 Write about what you would need to do in order to establish long-term temperature records for each month of the year.

Lesson 5 Time zones

While it is daytime in Papua New Guinea, it is night-time in some other parts of the world because as the Earth rotates different parts face the sun at different times. There are 24 hours in a day and each hour corresponds to one time zone, which is equivalent to 15° of longitude (15° × 24 hours = 360°). When you move from one time zone to the next you need to change your watch by one hour. If you travel in an easterly direction you move your watch one hour forwards; if you travel in a westerly direction you move your watch one hour backwards.

Help Box

The map below this box shows the world time zones. You can see that all of Papua New Guinea is in the same time zone. The prime meridian (longitude 0°) is the starting point from which time is measured either east or west. For example, to travel from Port Moresby to Los Angeles in the USA you cross 18 time zones (18 × 15°) in a westerly direction or 6 time zones (6 × 15°) in an easterly direction.

To find the time in Los Angeles you can either add 6 hours (one hour for each 15°) or you can subtract 18 hours (one hour for each 15°). If it is 3 p.m. in Port Moresby it will be 9 p.m. in Los Angeles: 3 p.m. + 6 hours = 9 p.m., or alternatively 3 p.m. – 18 hours = 9 p.m.

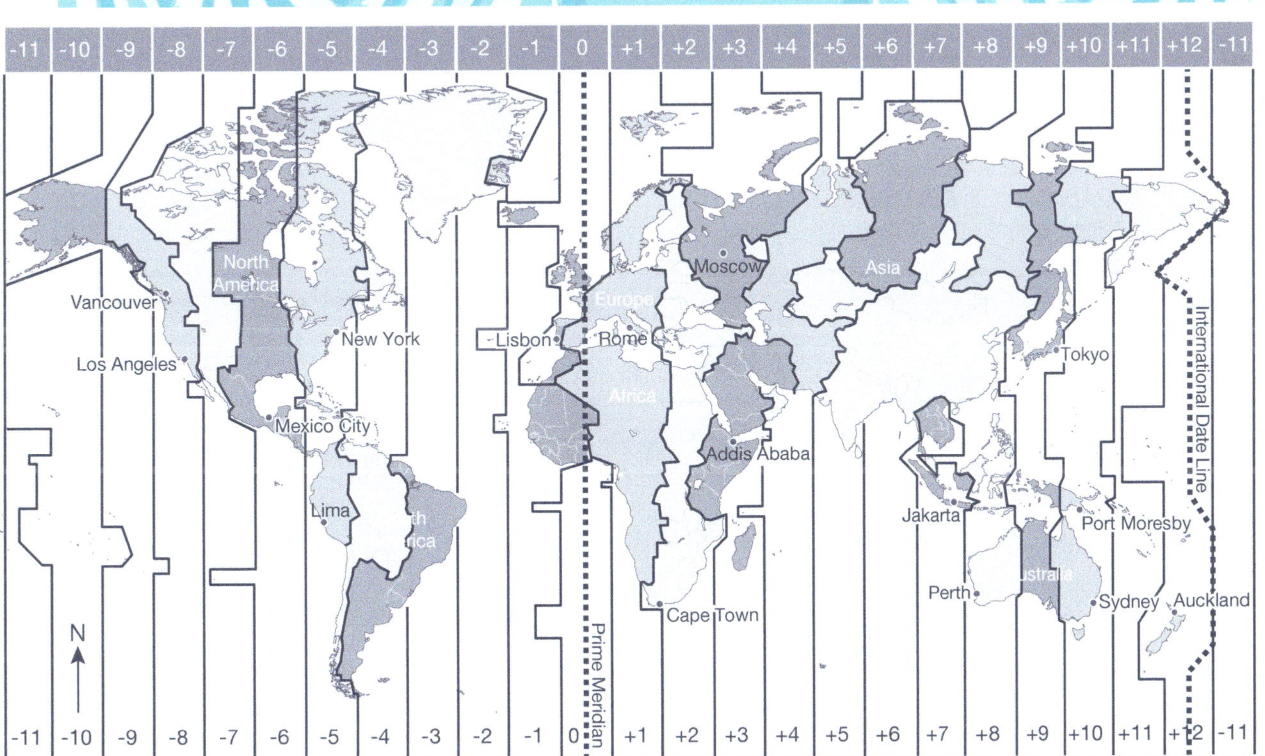

Use the time zones map to help you answer questions 1–8.

1 What is the time in the following places if it is 10 a.m. in Port Moresby?

a Sydney
b Perth
c Auckland
d Vancouver
e New York
f Tokyo
g Lima
h Mexico City

2 What is the time in the following places if it is 5 p.m. in Port Moresby?

a Cape Town
b Los Angeles
c Moscow
d Tokyo
e Jakarta
f Addis Ababa
g Lisbon
h New York

3 What is the time in the following places if it is 2 p.m. in Rome?

a New York
b Port Moresby
c Jakarta
d Cape Town
e Lima
f Lisbon
g Vancouver
h Perth

4 What is the time in the following places if it is midnight in New York?

a Moscow
b Tokyo
c Lisbon
d Rome
e Vancouver
f Port Moresby
g Perth
h Los Angeles

5 Ruth left Port Moresby on a 5 p.m. flight to Auckland in New Zealand.

a When it is 5 p.m. in Port Moresby, what is the time in Auckland?

b If the flight takes 6 hours, what is the local time in Auckland when Ruth lands?

c When Ruth lands in Auckland, what is the local time in Port Moresby?

6 Lionel, who lives in Madang, phoned a friend who lives in Perth.

a If the time was 10 a.m. in Madang, what time was it in Perth?

b If the phone call lasted 25 minutes, what was the time in Perth when the call ended?

7 Reuben left New York on a 9 p.m. flight to Tokyo in Japan.

a When it is 9 p.m. in New York, what time is it in Tokyo?

b If the flight takes 13.5 hours, what is the local time in Tokyo when Reuben lands?

c What time is it in New York when Reuben lands in Tokyo?

8 Lydia, who works in Port Moresby, is trying to organise a business phone conference between herself and two other people, one who lives in Jakarta and the other who lives in Vancouver.

a If Lydia organises the call for 9 a.m. Port Moresby time, what will be the time in Jakarta?

b What will the time be in Vancouver when it is 9 a.m. in Port Moresby?

c Do you think this is a good time to organise the phone conference? Why?

d Give two different times that would be suitable for the phone conference and say what the time would be in each of the three places.

The map below shows the part of the International Date Line that passes through the Pacific Ocean. It is an imaginary line that sits approximately on the 180° line of longitude.

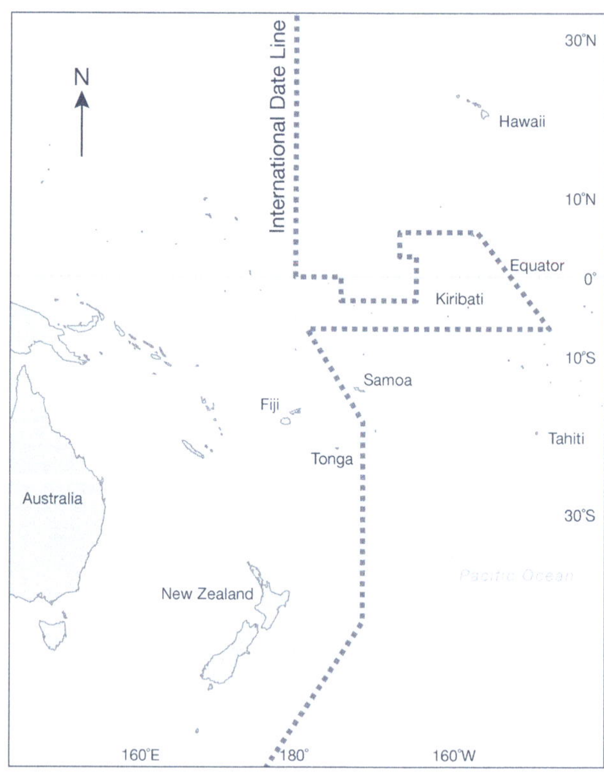

Help Box

The purpose of the International Date Line is to separate two calendar days or dates. If you cross the line in an easterly direction a day is subtracted. If you cross the line in a westerly direction a day is added. For example, if you leave Papua New Guinea on the morning of Sunday 16th and fly east to Tahiti, as soon as you cross the International Date Line it is Saturday 15th. On the return trip in a westerly direction, if you depart Tahiti on the morning of Thursday 20th as soon as you cross the International Date Line it is Friday 21st.

Tonga and Samoa have exactly the same time but because they are on opposite sides of the International Date Line they are one day apart. So, when it is Tuesday in Samoa, it is Wednesday in Tonga. As you can see on the map, the line bends around countries so all places in each country experience the same calendar day.

9 When it is Thursday lunchtime in Papua New Guinea, what day is it in each of the following places?

a Fiji
b Samoa
c Australia
d Hawaii
e Tonga
f New Zealand
g Kiribati
h Tahiti

10 When it is 10 a.m. in Port Moresby on 15 May, what time and date is it in each of the following places?

a Tokyo
b Vancouver
c Auckland
d New York

Challenge

Ruth's birthday is on 12 July. Last year she celebrated her birthday twice. Each celebration was on a different day but both days were 12 July. Explain how this could be possible.

Lesson 6 Plane specifications

All airline companies have to continually maintain and upgrade their fleet of aircraft. Aircraft need to be safe, reliable and cost effective.

If you travel overseas you could be a passenger on one of the aircraft shown in the table on the next page.

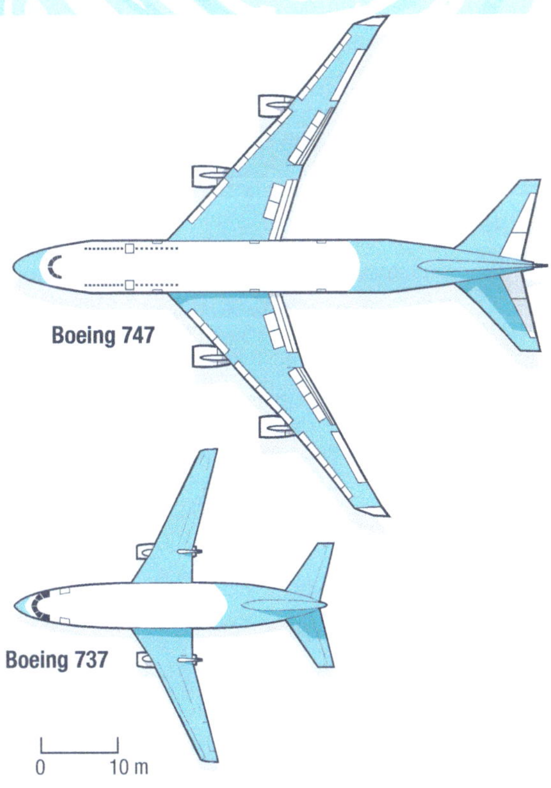

Name	Passengers	Fuel capacity	Wingspan	Length	Cruise speed	Range
Boeing 727	165	37 020 L	32.91 m	46.69 m	890 kph	4 020 km
Boeing 737	130	26 020 L	34.3 m	31.2 m	750 kph	5 650 km
Boeing 747	415	216 840 L	64.5 m	70.4 m	900 kph	13 450 km
Boeing 757	225	43 490 L	38.05 m	47.32 m	850 kph	7 220 km
Boeing 787	250	126 903 L	60 m	57 m	900 kph	14 200 km
Airbus 320	150	29 840 L	34.09 m	37.57 m	840 kph	4 800 km
Airbus 340	240	155 040 L	60.3 m	59.39 m	880 kph	14 800 km
Airbus 380	450	310 400 L	79.8 m	72.57 m	900 kph	15 200 km

The above table shows some of the important features of different aircraft. Use this information to help you answer the following questions.

1 Calculate the number of passengers on these aircraft if every seat was taken.

a 3 Boeing 747s
b 5 Airbus 340s
c 8 Boeing 737s
d 6 Boeing 727s
e 4 Airbus 320s and 5 Airbus 380s
f 7 Boeing 757s and 3 Boeing 787s

2 How many more passengers can an Airbus 380 carry than a Boeing 757?

3 How much longer is:

a an Airbus 340 than an Airbus 320
b a Boeing 787 than a Boeing 737
c an Airbus 380 than a Boeing 727
d a Boeing 747 than an Airbus 320?

4 What is the difference in the wingspan of:

a an Airbus 320 and a Boeing 787
b a Boeing 727 and an Airbus 380
c a Boeing 747 and a Boeing 757
d an Airbus 340 and a Boeing 727?

5 If five Airbus 380s were lined up one behind the other, what would be their total length?

6 How much fuel is needed to fill the following aircraft?

a 4 Airbus 320s
b 6 Boeing 747s
c 3 Boeing 787s
d 5 Airbus 340s
e 3 Boeing 757s and 2 Boeing 727s
f 4 Airbus 380s and 3 Airbus 340s

7 How long will it take the planes to travel the given distances?

a a Boeing 757 to travel 5100 km
b a Boeing 787 to travel 4680 km
c an Airbus 340 to travel 6820 km
d an Airbus 320 to travel 1890 km
e a Boeing 737 to travel 1050 km
f an Airbus 380 to travel 6600 km

8 Calculate the distance each of these planes would travel.

a a Boeing 737 in 2 h and 20 min

b an Airbus 340 in $5\frac{1}{2}$ h

c an Airbus 380 in 8 h and 45 min

d a Boeing 727 in $3\frac{1}{2}$ h

e a Boeing 757 in $4\frac{1}{4}$ h

f an Airbus 320 in $2\frac{3}{4}$ h

Help Box

To calculate the approximate fuel consumption per 100 km of any aircraft we can use the formula:

$$\text{Fuel consumption per 100 km} = \frac{\text{Fuel capacity}}{\text{Range}} \times 100$$

The range is the approximate distance a plane can travel when its fuel tank is full. The range can be affected by weather conditions and the weight the plane is carrying. For example, a Boeing 727 has a fuel capacity of 37 020 L and a range of 4020 km so:

$$\frac{37\,020}{4020} \times 100 = 9.21 \times 100$$
$$= 921 \text{ litres per 100 km}$$

9 Calculate the approximate fuel consumption per 100 km of these aircraft:

a Boeing 737

b Airbus 320

c Boeing 747

d Airbus 340

e Boeing 757

f Boeing 787

g Airbus 380

10 Compare the fuel consumption of the eight aircraft listed in the table.

a Which of the aircraft has the most efficient fuel consumption?

b Which has the least efficient fuel consumption?

Help Box

If we know the cruise speed and the range of an aircraft, we can calculate how long it can travel before it needs to refuel. We can use the formula:

$$\text{Flying time} = \frac{\text{Range}}{\text{Cruise speed}}$$

For example, a Boeing 727 has a range of 4020 km and a cruise speed of 890 kph.

$$\text{So, } \frac{4020}{890} = 4.52 \text{ hours}$$

To work out how many minutes 0.52 represents we say:

$0.52 \times 60 = 31.2$ or 31 minutes after rounding.

So, 4.52 hours is the same as 4 hours and 31 minutes.

The flying time of a Boeing 727 with a full fuel tank is 4 hours and 31 minutes.

11 Calculate the flying time for each of the aircraft in the table at the start of this lesson if their fuel tanks are full.

Use the information you calculated about flying time to answer questions 12–14.

12 Name all aircraft from the table that meet these requirements:

a can fly for at least 14 hours without refuelling

b could not fly for 6 hours without refuelling

c would need to stop once for refuelling on a 12-hour flight

13 You will need an atlas and a ruler for these questions.

a Find at least three places that an Airbus 320 could fly to from Port Moresby without having to stop to refuel.

b Starting from any major city in the world, find three flights that a Boeing 727 could make without stopping to refuel.

14 If you were making the following journeys, which aircraft would make each distance without having to stop to refuel and without having much fuel left over?

a Melbourne to Los Angeles
b Singapore to Perth
c Istanbul to Tokyo
d Hong Kong to Sydney
e London to Cairo
f San Francisco to New York

Challenge

Use the information in the aircraft table to make up at least five questions and answers of your own. Give them to a friend to solve.

Lesson 7 Pack your bag

Airlines place limits on the size and weight of bags to avoid overloading aircraft. Each plane has a maximum take-off weight above which it is not safe for the aircraft to take off. Passengers, luggage and fuel are just some of the items included in the maximum take-off weight. The more passengers on an aircraft, the less weight is available for other items.

Passengers are allowed two types of baggage: hand baggage that they take into the cabin and checked-in baggage that is loaded into the hold of the plane. Different airlines have different allowances. The following information shows the allowances for hand baggage in economy class for four airlines.

Airline 1 1 × 115 cm bag (maximum size) that does not exceed 7 kg
Airline 2 1 × 113 cm bag (maximum size) that does not exceed 7 kg
Airline 3 1 × 105 cm bag (maximum size) that does not exceed 7 kg
Airline 4 1 × 115 cm bag (maximum size) that does not exceed 8 kg

Help Box

Total baggage dimensions (or maximum size above) are calculated by adding together the length, height and depth of a bag. For example, a bag with a length of 56 cm, a height of 36 cm and a depth of 23 cm has a total measurement of 115 cm:

56 cm + 36 cm + 23 cm = 115 cm.

1 What other dimensions could a bag have so that its total is no more than 115 cm? Give at least three realistic possibilities.

2 Would Airline 3 accept bags of the following dimensions?

a 53 cm, 37 cm and 22 cm
b 49 cm, 32 cm and 20 cm
c 45 cm, 34 cm and 25 cm

3 Would Airline 2 accept bags of the following dimensions?

a 52 cm, 36 cm and 24 cm

b 55 cm, 45 cm and 18 cm

c 49 cm, 38 cm and 23 cm

4 Choose items that you estimate have a combined mass of about 7 kg. Weigh the items to see how close your estimate was and either add or take away items until you have as near to 7 kg as possible.

Different airlines also vary in their allowances for checked-in baggage. The table below shows the allowances for checked-in baggage in first class, business class and economy class for two airlines.

	First class	Business class	Economy class
Airline A	40 kg	30 kg	20 kg
Airline B	32 kg × 2 bags	32 kg × 2 bags	23 kg × 2 bags

5 How much more weight are you allowed in each class with Airline B than Airline A?

6 187 checked-in bags for passengers flying economy class are loaded onto an aircraft. The mean weight of each bag is 19.65 kg.

a What is the total weight of the 187 bags?

b How much less than 5000 kg is this?

7 A 225-seater plane is full. It carries 35 first class passengers, 52 business class passengers and the rest are economy passengers.

a How many passengers are in economy class?

b If each passenger checked in the maximum allowable baggage according to Airline A above, what weight is loaded on board?

c If each passenger checked in the maximum allowable baggage according to Airline B above, what weight is loaded on board?

8 The total mass of baggage checked in for 27 first class passengers is 1014.66 kg, the total mass for 43 business class passengers is 1197.55 kg and the total mass for 96 economy class passengers is 1763.52 kg.

a Calculate the combined mass for the three classes.

b What is the mean mass of a first class passenger's bag?

c What is the mean mass of a business class passenger's bag?

d What is the mean mass of an economy class passenger's bag?

Challenge

Find a bag and pack it with items such as clothing and shoes. When you think it weighs 20 kg, check to see how close you are. Remove or add items to achieve a mass as close as possible to 20 kg.

Sometimes travellers need to take more luggage than is allowed. This extra luggage is known as 'excess baggage' and passengers have to pay additional charges to transport it. Some airlines charge per kilogram while others charge per piece of luggage.

9 An airline charges K24.75 per kilogram for excess baggage. How much will it cost for excess baggage of the following masses?

a 15 kg   b 9.5 kg   c 23 kg

d 18.7 kg   e 29.4 kg

10 An airline charges K42.50 per kilogram for excess baggage. How much will it cost for excess baggage of the following masses?

a 7.6 kg   b 12.3 kg   c 9.8 kg

d 16.25 kg   e 21.75 kg

11 A passenger pays a total of K542.50 for excess baggage. If the rate is K35 per kilogram, how many kilograms of extra baggage does the passenger have?

12 A passenger pays a total of K733.60 for excess baggage. If the passenger has 22.4 kg of extra baggage, how much is he paying per kilogram?

13 Jonah's bag weighed 32.5 kg. He was allowed 20 kg and then had to pay K26.75 per kilogram for excess baggage. How much did Jonah pay for excess baggage?

14 Antonia was allowed 30 kg of luggage. She had 37.5 kg. If she is charged K412.50 for her excess baggage, how much is she paying per kilogram?

Lesson 8 What is my money worth?

People visiting Papua New Guinea have to use kina and toea when buying goods and services. Travellers to other countries have to use the currency of the country they are visiting. Travellers can exchange currency at most airports or banks. The exchange rates (also called the foreign exchange rate, forex rate or FX rate) between two currencies states how much one currency is worth in terms of the other. Many factors influence exchange rates, causing them to fluctuate from day to day.

A group of students investigated the exchange rate the week before they were to travel overseas and made the following list to show what they would get in exchange for K1 in the following currencies.

Currency	Exchange rate	Currency	Exchange rate
Thailand baht	11.94 baht	US dollars	0.37 dollars
Australian dollars	0.40 dollars	Fiji dollars	0.56 dollars
Singapore dollars	0.51 dollars	European euros	0.24 euros
Japanese yen	38 yen	South Africa rand	2.89 rand
English pounds	0.19 pounds	Philippines pesos	15.89 pesos

1 How many Australian dollars would they get for these amounts?

a K100 b K150 c K300
d K250 e K500

2 How many Thailand baht would they get for these amounts?

a K75 b K90 c K200
d K125 e K240

3 How many Japanese yen would they get for these amounts?

a K50 b K120 c K75
d K375 e K180

4 How many euros would they get for these amounts?

a K175 b K65 c K120
d K275 e K235

5 If the students changed K250 into each of these currencies, how much of each would they get?

a Singapore dollars b English pounds
c US dollars d Fiji dollars
e South African rand f Philippines pesos

6 If the students changed K475 into each of these currencies, how much of each would they get?

a US dollars b Japanese yen
c English pounds d South African rand
e Australian dollars f Fiji dollars

7 Eddie was travelling to Melbourne in Australia and exchanged K2000 for Australian dollars at the rate shown in the list above.

a How many Australian dollars did Eddie get?

b How many Australian dollars would Eddie get if the exchange rate was K1 = 0.42 dollars?

8 Josie was travelling to London in England. She exchanged K1500 for English pounds at a rate of 0.20 pounds per kina.

a How many English pounds did Josie get?

b How much more money did she get than if the exchange rate was as shown in the list above?

c How many English pounds would Josie get if the exchange rate was K1 = 0.22 English pounds?

9 Cecilie was travelling to Singapore. She exchanged K500 for Singapore dollars at a rate of 0.49 dollars per kina.

a How many Singapore dollars did Cecilie get?

b How much less money did she get than if the exchange rate was as shown in the list above?

c How many Singapore dollars would Cecilie get if the exchange rate was K1 = 0.52 Singapore dollars?

10 Wesley was travelling to Manila in the Philippines and exchanged K750 for Philippines pesos at the rate shown in the list above.

a How many Philippines pesos did Wesley get?

b How many Philippines pesos would Wesley get if the exchange rate was K1 = 15.9 pesos?

11 Sumasi was travelling to Cape Town in South Africa and exchanged K1200 for South African rand at the rate shown in the list above.

a How many South African rand did Sumasi get?

b How many South African rand would Sumasi get if the exchange rate was K1 = 2.85 rand?

Help Box

When travellers return to Papua New Guinea they need to exchange their foreign money for kina. To calculate the amount they receive they can use the following process:

$$\text{kina} = \frac{\text{foreign money}}{\text{exchange rate}}$$

For example, if Eddie had 200 Australian dollars left and he used the exchange rate for kina and Australian dollars in the list, the calculation would be:

$$\frac{200}{0.4} = 500$$

So Eddie would get K500.

Use the exchange rates shown in the earlier list to answer questions 12–15.

12 Calculate how many kina each of the following amounts would be exchanged for.

a 200 US dollars
b 120 Australian dollars
c 150 Fiji dollars
d 800 Philippines pesos
e 420 South African rand
f 60 euros

13 When Josie returned to Papua New Guinea she had 100 English pounds left which she exchanged for kina. How many kina did she get?

14 When Cecilie returned to Papua New Guinea she had 60 Singapore dollars left which she exchanged for kina. How many kina did she get?

15 Thomas had been in Japan and when he returned he had 1000 Japanese yen to exchange for kina. How many kina will he get?

16 When Martin went to New York in the United States of America the exchange rate was K1 = 0.37 US dollars. On his return he found that the exchange rate was now K1 = 0.39 US dollars.

a If Martin exchanges 500 US dollars, how many kina will he get?

b If the exchange rate was the same as when Martin left, how many kina would he have got?

c What is the difference between the two amounts of money?

17 When Alice went to Bangkok in Thailand the exchange rate was K1 = 11.94 baht. On her return she found that the exchange rate was now K1 = 12.08 baht.

a If Alice exchanges her 2000 Thailand baht, how many kina will she get?

b If the exchange rate was the same as when Alice left, how many kina would she have got?

c What is the difference between the two amounts of money?

18 Jerry exchanged 200 euros at a rate of K1 = 0.24 euros. Arnold exchanged the same amount of euros at a rate of 0.26 euros.

a Which student would get more kina?

b What is the difference between the two amounts of money?

Challenge

Ask your teacher to help you find the exchange rates for the currencies of two or three countries different to the ones mentioned in this lesson. Calculate how much of each currency K500 buys.

Lesson 9 Airfare deals

The table below shows a selection of economy and business class return fares between Port Moresby and ten cities around the world. Airfares for the same route can vary considerably so it pays to check all available fares before buying.

City	Economy class	Business class
Brisbane	K3 124.10	K8 252.49
Singapore	K11 524.21	K21 858.43
Hong Kong	K7 443.31	K20 419.42
Tokyo	K6 612.08	K27 177.54
Auckland	K5 996.65	K11 512.14
London	K10 553.68	K32 670.83
Los Angeles	K6 781.87	K43 910.41
Paris	K16 131.06	K39 891.63
Nadi (Fiji)	K6 076.15	K8 426.83
Beijing	K9 750.60	K21 764.34

Use the fares in the table to answer questions 1–6.

1 Find the difference in price between an economy class fare and a business class fare for a return trip between Port Moresby and these cities.

a London b Singapore c Paris

d Beijing e Auckland

2 Which destination has the greatest price difference between economy class and business class?

3 Calculate the cost of the following return tickets.

a 2 economy class to London

b 5 business class to Nadi

c 3 business class to Tokyo

d 4 economy class to Brisbane

e 2 economy class and 1 business class to Los Angeles

f 3 economy class and 2 business class to Hong Kong

4 How much more does it cost to travel economy class between:

a Port Moresby and Singapore than between Port Moresby and Hong Kong?

b Port Moresby and Los Angeles than between Port Moresby and Auckland?

c Port Moresby and Paris than between Port Moresby and London?

Many international airlines offer discounts for children travelling with adults. Children under 2 years of age are generally charged 10% of the adult economy fare and are expected to sit on an adult's lap. Children between the ages of 2 and 11 are generally charged 75% of the adult economy fare and have their own seat.

5 Calculate the cost of return economy fares at the above discounts departing from Port Moresby for these families.

a Two adults and two children under the age of 2 flying to London.

b One adult and one 8-year-old child flying to Los Angeles.

c Three adults and two children between 2 and 11 flying to Auckland.

d Two adults, one child under the age of 2 and one 5-year-old flying to Hong Kong.

e One adult and two children between 2 and 11 flying to Tokyo.

f Two adults and three children between 2 and 11 flying to Paris.

6 If an airline allowed children between the ages of 2 and 11 a discount of 33% off the adult economy fare, calculate the cost of economy fares for these families.

a Two adults and one 10-year-old flying to Nadi.

b One adult and one 6-year-old flying to Singapore.

c Two adults and two children between 2 and 11 flying to Beijing.

d Two adults and one 7-year-old flying to Brisbane.

e Two adults, one child under the age of 2 and one 4-year-old flying to London.

f Three adults and two children between 2 and 11 flying to Tokyo.

Help Box

Sometimes you will be given the percentage of a particular amount and then asked to work out the full amount. For example, Charles was charged K4650 which was 75% of the adult fare. What is the full adult fare?

Examples of this type can be worked out in the following way.

Step 1: Find 1% of the amount by dividing K4650 by 75
$4650 \div 75 = K62$

Step 2: Find 100% of the amount by multiplying K62 by 100
$K62 \times 100 = K6200$

The original amount was K6200.

7 The following fares represent 75% of the full adult fare. Calculate each adult fare.

a K3675 b K4275 c K6225
d K7275 e K5400 f K4743.75
g K3800.25 h K3112.50 i K4303.50
j K1622.25

8 The following fares represent 67% of the full adult fare. Calculate each adult fare.

a K2412 b K5427 c K3618
d K4824 e K4221 f K2587.54
g K1284.39 h K5605.22 i K4976.76
j K3965.73

9 Daniel paid K2156 for a return ticket from Port Moresby to Brisbane. He received 30% off the full fare. How much was the full fare?

10 When Patricia flew to Auckland she received a 5% discount off the full return fare. If she paid K5543.25, how much was the full fare?

11 Jane paid 85% of the adult return fare to Tokyo. She paid a total of K5673.75. How much was the adult return fare?

12 Stephen paid 80% of the adult return fare from Port Moresby to Singapore. If Stephen paid K8765.60, how much was the adult fare?

13 Copy this table into your workbook and complete the missing fares.

	Full fare	30% off full fare	25% off full fare	10% off full fare
a	K7543			
b		K3446.10		
c				K4647.60
d			K2581.50	
e		K1899.10		
f	K9715			
g				K5704.20
h			K4499.25	
i	K4289			
j		K5874.40		

Learning Unit 3 Tourism Comparisons

Strand: Number and Application

Decimals	Outcome 8.1.2	Use decimals to solve real life problems
Fractions and Decimals	Outcome 8.1.3	Convert freely between fractions, decimals, percentages and ratios
Decimals and Percentages	Outcome 8.1.4	Solve problems in any situation that involves percentages

Strand: Space and Shape

Capacity	Outcome 8.2.8	Convert between a variety of units: capacity and volume, and metric and imperial

Strand: Chance and Data

Statistics	Outcome 8.4.1	Interpret information presented statistically
Accuracy and Error	Outcome 8.4.5	Represent levels of accuracy
Estimation	Outcome 8.4.7	Estimate results of calculations

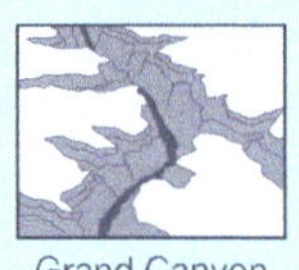
Grand Canyon

Statue of Liberty

Tower of London

Eiffel Tower

Leaning Tower of Pisa

Lesson 1: Introduction	Locating major attractions around the world Comparing visitor numbers at major attractions
Lesson 2: Price comparisons around the world	Converting between kina and other currencies Comparing and calculating with decimals
Lesson 3: Physical comparisons	Calculating with decimals Converting between metres and feet Converting between kilometres and miles
Lesson 4: Where the people go	Interpreting and drawing comparative bar graphs Interpreting, comparing and drawing line graphs Calculating percentage increases and decreases
Lesson 5: Arrivals in Papua New Guinea	Interpreting and drawing bar graphs, line graphs and pie charts Writing tourism statistics as percentages Calculating percentage increases and decreases Increasing and decreasing by a given percentage
Lesson 6: Cost of accommodation around the world	Using decimals and percentages to solve problems Converting between decimals and percentages Increasing and decreasing by a given percentage
Lesson 7: Cargo and passenger statistics	Calculating the mean Calculating percentage increases and decreases Increasing and decreasing by a given percentage
Lesson 8: Some interesting comparisons	Using decimals in calculations Writing data as fractions and decimals

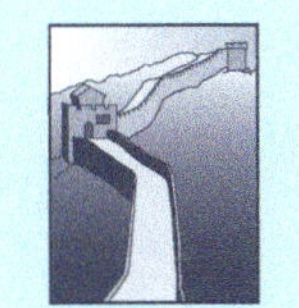
Great Wall of China

Disneyland

Uluru

Taj Mahal

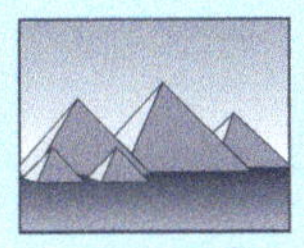
Pyramids of Giza

Lesson 1 Introduction

Just as tourists come to Papua New Guinea to experience the many attractions that are on offer, so we travel to other countries to see their attractions. During this unit you will compare some of the facts and figures that are associated with the tourism industry in different countries.

The map and illustrations below show some of the major attractions that tourists visit around the world.

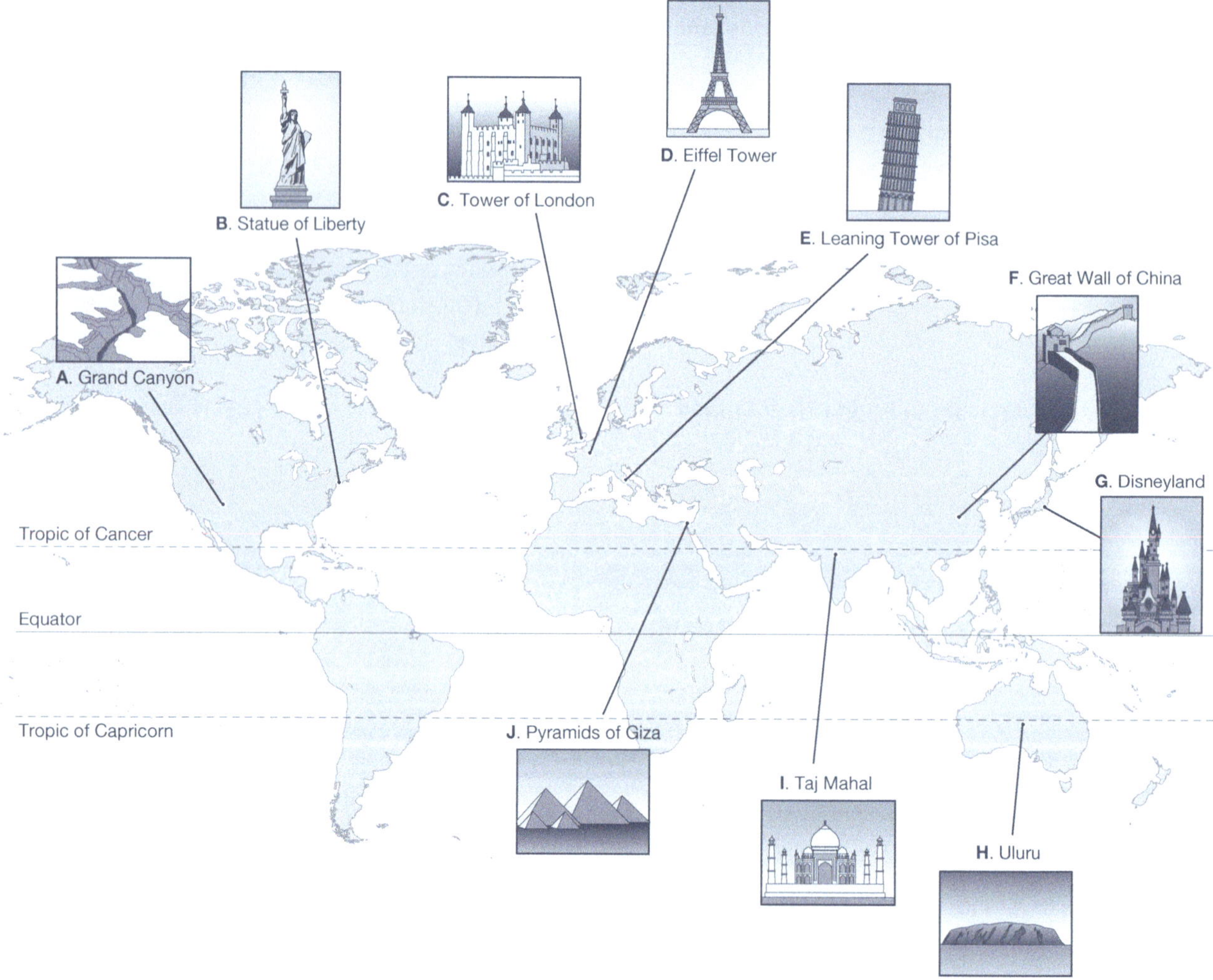

1 In which country is each of the attractions shown on the previous page? Use an atlas to help you find the name of each country and make a list (A to J) in your book.

2 Which of the ten attractions would you most like to visit? Why?

3 This chart shows the approximate number of people who visit each of the attractions annually.

Tower of London	2 500 000	Great Wall of China	10 000 000
Eiffel Tower	6 700 000	Uluru	400 000
Disneyland (Tokyo)	12 900 000	Taj Mahal	2 400 000
Grand Canyon	4 400 000	Pyramids of Giza	3 000 000
Statue of Liberty	4 240 000	Leaning Tower of Pisa	1 200 000

a Which of the attractions has the most annual visitors?

b Which has the least number of visitors annually?

c Put the attractions in order from most annual visitors to least annual visitors.

Lesson 2 Price comparisons around the world

Comparing the cost of items in various countries or cities around the world helps us to gain an idea of the cost of living in those places.

The table below shows the cost of a hamburger in different countries and the comparative cost when all prices are converted to US dollars.

Country	Price	Cost in US dollars	Cost in kina
USA	US$3.41	$3.41	K9.17
Australia	AUS$3.45	$3.30	K8.88
England	1.99 pounds	$3.90	
China	yuan 11	$1.60	
Denmark	dkr 27.75	$5.78	
France	3.06 euros	$4.76	
Hong Kong	HK$12	$1.54	
Indonesia	rupiah 15 900	$1.72	
Japan	yen 280	$2.69	
Norway	kroner 40	$7.94	
Philippines	peso 85	$1.98	
Singapore	S$3.95	$2.89	

1 Copy the table into your workbook and use the exchange rate US$1 = K2.69 to convert each US dollar amount to kina. The first two have been done for you.

2 Find the difference in kina of the price of a hamburger between these countries.

a Australia and Norway
b USA and Singapore
c Denmark and England

3 In which country is a hamburger:

a the most expensive
b the least expensive?

Challenge

Find the current exchange rate and, if it is different to the one used above to convert US dollars to kina, use it to find the cost of a hamburger in each country listed in the table.

The following table shows price comparisons in euros of some common items in five major cities.

	New York	London	Paris	Tokyo	Dublin
loaf of bread	€1.90	€1.40	€2.10	€3.90	€1.40
2 kg bag potatoes	€1.45	€2.75	€2.50	€4.75	€2.60
cup of coffee	€2.85	€2.95	€3.50	€3.45	€3.00
bus/tube ticket	€1.52	€4.45	€1.40	€1.95	€1.50
CD (music)	€13.25	€19.15	€19.00	€13.50	€20.00
The symbol € stands for euros.					

4 Rule a table in your workbook like the one above but, instead of converting to euros, convert each amount to kina using the exchange rate €1 = K4.18.

5 Find the price of each item listed in the table in your local area in Papua New Guinea and enter them in a new column beside the other five columns.

6 Use the six sets of prices (in kina) in your table to answer the following questions.

a How much more expensive is the most expensive cup of coffee than the cheapest cup of coffee?

b How much more will a CD cost in London than in New York?

c What is the price difference between 2 kg of potatoes in Papua New Guinea and in Dublin?

d Order the places from most expensive to least expensive for a loaf of bread.

7 Imagine that you bought one of each item from each place listed in your table. Find the total cost for the five items in each city and rank the cities in order from most to least expensive.

8 Based on the prices in your table, what would the following items cost?

a 3 × 2 kg bags of potatoes in Tokyo
b 5 cups of coffee in London
c 6 bus tickets in New York
d 4 CDs in Paris
e 3 loaves of bread in Tokyo
f 3 cups of coffee in your local area

The results of a survey based on the cost of a 2GB iPod Nano in US dollars are shown below. It is most expensive in the first five countries and least expensive in the second five countries.

Country	Cost in US$
Brazil	$327.71
India	$222.27
Sweden	$213.03
Denmark	$208.25
Belgium	$205.81
Mexico	$154.46
United States	$149.00
Japan	$147.63
Hong Kong	$147.35
Canada	$144.20

9 A 2GB iPod Nano costs K685 in Papua New Guinea.

a Convert K685 to US dollars using the exchange rate K1 = US$0.37.

b Copy the above list into your workbook and enter Papua New Guinea in its correct place compared to the other countries.

c How would you describe the cost of a 2GB iPod Nano in Papua New Guinea compared to other countries?

10 Compare the price differences between the following countries.

a How much more does an iPod cost in Brazil than in Canada?

b How much cheaper is an iPod in Japan than in Papua New Guinea?

c How much more would you pay for an iPod in Sweden than in Mexico?

One of the main expenses when travelling is the cost of entry to places and attractions. Some examples are given below.

	Adult	Child
Tour of Sydney Opera House (Australia)	A$16.20	A$11.10
Great Wall of China (China)	¥100	¥100
Eiffel Tower (France)	€9.90	€5.30
Disneyland Paris (France)	€36	€29
Colosseum in Rome (Italy)	€8	€8
Kruger National Park (South Africa)	R30	R15
Serengeti National Park (Africa)	US$30	US$5
Tower of London (England)	£11.50	£7.50
Statue of Liberty (United States)	US$8	US$3
Disneyland California (United States)	US$43	US$33

This table shows how much K1 buys in each currency above.

0.40 Australian dollars (A$)	2.80 South African rand (R)
2.59 Chinese yuan (¥)	$0.37 United States dollars (US$)
0.24 euros (€)	0.19 English pounds (£)

Help Box

In Lesson 8 in the previous unit, you learned how to convert other currencies to kina using the formula

$$\text{kina (K)} = \frac{\text{foreign money}}{\text{exchange rate}}$$

The entry fees shown above can be converted to kina so that we know how many kina we need to budget for, and so we can compare the entry costs to similar attractions and decide which to visit.

Example: To calculate the entry fee to the Great Wall of China we say:

$$\frac{¥100}{2.59} = K38.61.$$

11 Convert the following entry fees to kina.

a Eiffel Tower – Adult
b Kruger National Park – Child
c Disneyland Paris – Adult
d Statue of Liberty – Child
e Tower of London – Child
f Colosseum – Adult

12 Calculate the following entry fees in kina.

a 2 adults and 2 children to the Tower of London
b 3 adults to the Great Wall of China
c 2 adults and 1 child to the Sydney Opera House
d 6 children to Disneyland in California
e 1 adult and 4 children to Disneyland in Paris
f 4 adults to Serengeti National Park

13 How much more does it cost for 2 adults and 2 children to enter Disneyland in Paris than Disneyland in California?

14 How much more does it cost for 3 adults to enter Serengeti National Park than Kruger National Park?

15 On a trip to England and France a family of five, made up of 2 adults and 3 children, visited the Tower of London, the Eiffel Tower and Disneyland Paris. How much in total did they spend on entry fees to the three places?

Challenge

Use the information about entry fees to make up two questions of your own similar to the ones you have done. Give them to a friend to solve.

Lesson 3 Physical comparisons

Some of the most visited attractions around the world are buildings and towers. The list below shows some of the most popular buildings and towers and the height of each structure, including any antennas or spires, in metres.

Building or Tower	Height (metres)	Building or Tower	Height (metres)
Notre Dame Cathedral (Paris, France)	90	Petronas Twin Towers (Kuala Lumpur, Malaysia)	452
Eiffel Tower (Paris, France)	324	CN Tower (Toronto, Canada)	553.3
Empire State Building (New York, USA)	448.7	Eureka Tower (Melbourne, Australia)	297.3
Chrysler Building (New York, USA)	318.9	Taipei 101 (Taipei, Taiwan)	509.2
Sears Tower (Chicago, USA)	527.3	Burj Al Arab Hotel (Dubai, United Arab Emirates)	321

1 How much taller is:

a Sears Tower than the Chrysler Building?

b Taipei 101 than the Empire State Building?

c the Burj Al Arab Hotel than Eureka Tower?

d CN Tower than the Eiffel Tower?

e Empire State Building than the Chrysler Building?

f Petronas Twin Towers than Eureka Tower?

2 What is the height difference between the tallest structure and the shortest structure in the list above?

Help Box

Most countries use the metric system for length, and give measurements in kilometres, metres, centimetres and millimetres. However, some countries still use the imperial system and give measurements in miles, yards, feet and inches. In some countries a combination of both systems is still accepted and used.

We can convert between metres and feet by using the rate: 1 metre = 3.28 feet.

For example, to convert 90 metres to feet we calculate $90 \times 3.28 = 295.2$ feet.

To convert feet to metres we reverse the procedure: $295.2 \div 3.28 = 90$ metres.

When converting feet to metres, we can also use 1 foot $\approx$ 0.3 metres to multiply, so that $295.2 \times 0.3 \approx 88.56$ metres which is close to 90 metres.

3 Convert the following lengths to feet. Round your answers to the nearest tenth.

a 20 m b 100 m c 45 m

d 58 m e 74 m f 67 m

4 Convert the height in metres of each structure in the list of buildings to feet, and then answer these questions.

a Which structures are taller than 1500 feet?

b Which structures are between 900 and 1100 feet in height?

c How much less than 2000 feet is the tallest structure?

5 Convert the following lengths to metres by dividing. Round your answers to the nearest tenth. (If you have a calculator you may use it.)

a 82 feet b 328 feet c 150 feet

d 430 feet e 275 feet f 186 feet

6 Convert the lengths in question 5 to metres by multiplying by 0.3. Compare your two sets of answers.

7 The tallest mountain in the world is Mt Everest with a height of 8852 metres. Write this height in feet to the nearest whole foot.

8 The table below shows the height of some other tall mountains. Copy it into your book and complete the gaps. Round all heights to the nearest metre or foot.

Mountain	Height (m)	Height (ft)
K2	8612	
Kangchenjunga	8586	
Lhotse		27 883
Makalu		27 755
Cho Oyu	8201	

9 The highest mountain in Papua New Guinea is Mount Wilhelm at a height of 14 790 feet. Estimate and then calculate its height to the nearest whole metre.

Help Box

Longer lengths and distances are usually given in kilometres (metric) or miles (imperial).

We can convert between kilometres and miles by using the rate: 1 kilometre = 0.62 miles

For example, to convert 75 kilometres to miles we calculate $75 \times 0.62 = 46.5$ miles.

To convert miles to kilometres we reverse the procedure: $46.5 \div 0.62 = 75$ kilometres.

When converting miles to kilometres we can also use 1 mile ≈ 1.6 kilometres to multiply, so that $46.5 \times 1.6 \approx 74.4$ kilometres which is close to 75 kilometres.

10 Convert the following lengths to miles. Round your answers to the nearest tenth.

a 80 km b 120 km c 35 km d 97 km e 152 km f 233 km

11 Convert the following lengths to kilometres by dividing. Round your answers to the nearest tenth. (If you have a calculator you may use it.)

a 65 miles b 78 miles c 115 miles d 223 miles e 146 miles f 287 miles

12 Convert the lengths in question 11 to kilometres by multiplying by 1.6. Compare your two sets of answers.

13 The longest river in Papua New Guinea is the Sepik with a length of 1126 kilometres. Write this length in miles to the nearest whole mile.

14 The table below shows the length of some of the longest rivers in the world. Copy it into your book and complete the gaps. Round all lengths to the nearest kilometre or mile.

River	Length (km)	Length (miles)
Nile		4160
Amazon	6452	
Yangtze	6394	
Congo		2718
Mekong	4194	
Mississippi	3742	

15 The Lake Pontchartrain Causeway across Lake Pontchartrain in the United States of America consists of two parallel bridges. The longer of the two bridges is 38.5 kilometres long. Write this length in miles to two decimal places.

16 Hangzhou Bay Bridge in China is 22.37 miles long. Donghai Bridge in China is 32.5 kilometres long. Which bridge is longer and by how much?

17 To cross the Vasco da Gama Bridge in Portugal you travel 10.66 miles. What is this distance in kilometres to two decimal places?

Lesson 4 Where the people go

Every year people visit other countries for a variety of reasons that include holidays, business trips and visiting relatives or friends.

This graph shows the number of people from other countries that visited various Pacific countries in 2002 and 2003.

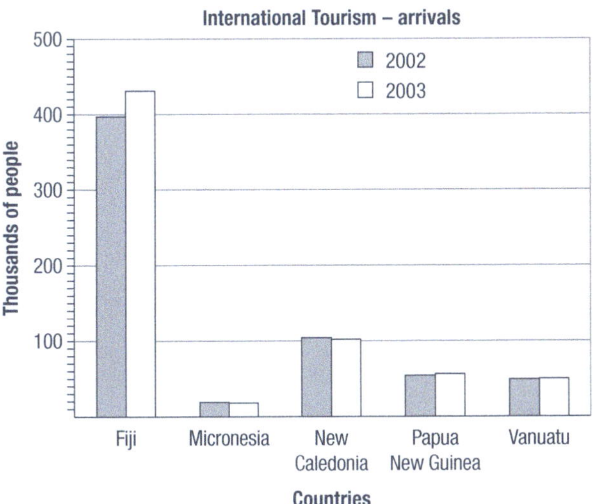

1 Which country had:

a the most visitors in 2002

b the least visitors in 2003

c the most visitors in any year?

2 Write the approximate number of visitors to the following countries:

a Papua New Guinea in 2003

b New Caledonia in 2002

c Vanuatu in 2003

3 Which countries had an increase in visitors from 2002 to 2003?

You can see that in 2002 Fiji had almost four times the number of visitors that New Caledonia had.

4 For the year 2003 describe:

a the number of visitors to Fiji in comparison to the number of visitors to Papua New Guinea

b the number of visitors to Vanuatu in comparison to the number of visitors to Micronesia

The previous graph is difficult to read accurately because of the large scale on the *y*-axis caused by the inclusion of the arrivals data from Fiji. The same data, excluding Fiji, is represented on the graph below.

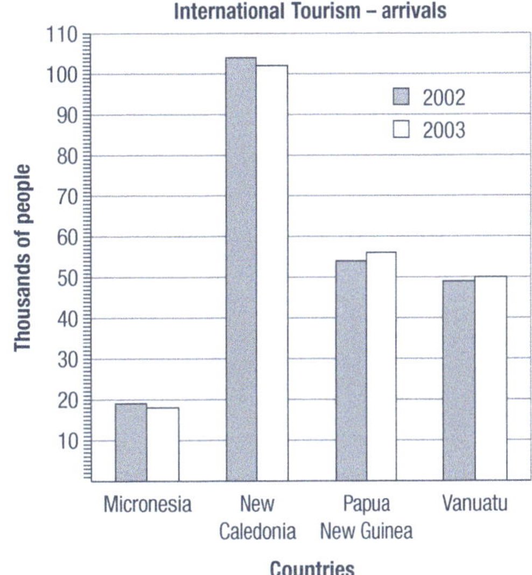

5 In 2003 which countries had the following number of visitors?

a 56 000 b 102 000

c 50 000 d 18 000

6 In 2002 how many more visitors did New Caledonia have than the following countries?

a Vanuatu b Papua New Guinea

c Micronesia

7 How many more visitors in total did Papua New Guinea have than Micronesia for 2002 and 2003?

8 For the year 2003:

a Describe the number of visitors to Vanuatu in comparison to the number of visitors to Micronesia.

b Compare your description with your answer to question 4b.

c Which description is more accurate? Why?

The table below shows the number of visitors rounded to the nearest thousand for the years 2004 and 2005 to the same four countries as shown in the previous graph.

	2004	2005
Papua New Guinea	61 000	68 000
New Caledonia	100 000	101 000
Micronesia	19 000	19 000
Vanuatu	61 000	62 000

9 Draw a graph similar to the previous one to display the data contained in the table. Make sure you choose a suitable scale, label the axes, give your graph a title and use a key to distinguish 2004 data from 2005 data.

10 Use your graph to answer the following questions:

a Which countries had an increase in visitors from 2004 to 2005?

b Which country had the greatest visitor increase from 2004 to 2005?

c How many more visitors in total did Papua New Guinea have than Vanuatu for 2004 and 2005?

These line graphs show tourism arrivals from 1996 to 2005 in four countries.

International Tourism—Arrivals

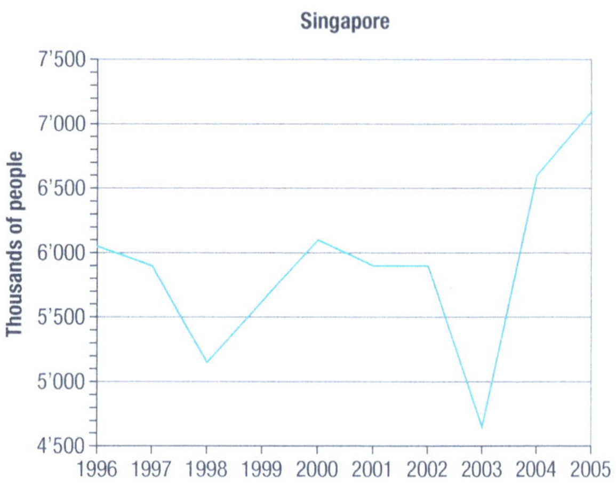

You will notice that the scale on the y-axis on each graph is in thousands of people, so that 4'800, for example, means 4 800 000.

Look at the line graphs to help you answer questions 11 to 17.

11 Which country had the most arrivals in 1996?

12 In 2003 which countries had the following number of visitors?

a 4 700 000 b 4 475 000

c 2 110 000 d 4 750 000

13 Between which two years is there a reduction in tourist numbers to each of the four countries? What might be the cause of this reduction?

14 Three of the graphs show a reduction in tourist numbers between 2002 and 2003.

a Which countries had a reduction in tourist numbers between these years?

b Which of the three countries had the largest reduction?

c Approximately how many less tourists visited each of the three countries in 2003 than in 2002?

d Suggest reasons for the reduction in visitors to these countries.

15 Calculate the percentage increase or decrease in tourist arrivals for the countries and years listed below. (If you have forgotten how to do this refer to Lesson 10 in Learning Unit 1.) Write the percentage decreases with a minus sign in front of them, for example –20%.

a Singapore from 2002 to 2003

b Indonesia from 1998 to 2001

c Australia from 2000 to 2003

d New Zealand from 1999 to 2002

e Indonesia from 1998 to 2000

f Australia from 2003 to 2005

16 Use the graphs to predict to the nearest hundred thousand the tourist arrivals for 2006 in:

a New Zealand b Australia

c Singapore

17 Plot the data for Australia and Indonesia on the same graph. You will need to choose a suitable scale and use a different coloured line for each country.

18 Write five statements about your completed graph that compare the tourist numbers in each country.

Lesson 5 Arrivals in Papua New Guinea

This bar graph shows the number of people (rounded to the nearest thousand) that visited Papua New Guinea from overseas countries each year from 2000 to 2007.

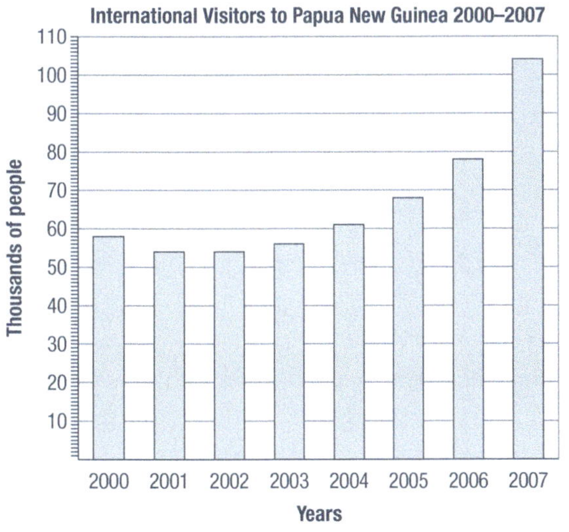

1 Use the graph to help you answer these questions.

a Which year had the most international visitors?

b Which years had the least international visitors?

c How many more visitors were there in 2006 than in 2001?

d How many more visitors were there in 2007 than in 2000?

2 Calculate the percentage increase in visitor arrivals from:

a 2005 to 2006 b 2006 to 2007

c 2003 to 2007

3 Draw a line graph to display the same data shown on the bar graph. Use your line graph to predict the visitor arrival numbers for Papua New Guinea in 2008.

Visitors come to Papua New Guinea from many different countries. Many of those visitors are from Australia, Japan and the United States of America. You can see from the table below that Australian visitors comprise approximately half of the annual visitors.

Total visitors to Papua New Guinea				
Year	**Total visitors (from all countries)**	**From Australia**	**From USA**	**From Japan**
2000	58 448	29 285	5429	3244
2001	54 235	27 661	5314	2686
2002	53 762	26 650	6053	3804
2003	56 282	30 118	4566	3893
2004	60 715	33 255	4822	3605
2005	68 450	36 401	5186	5397
2006	77 730	40 642	6228	3966
2007	104 122	54 098	6159	3347

Use the information in the table to help you answer the following questions.

4 In what years was the percentage of visitors from Australia more than 50% of the total visitors?

5 Calculate the percentage of the total number of visitors from:

a Japan in 2007 b the USA in 2007 c the USA in 2006 d Japan in 2006

6 Calculate the percentage increase or decrease in the number of visitors from each country from 2006 to 2007. Make sure you indicate if it is an increase or decrease for each country.

7 This pie chart represents the percentage of visitor arrivals (rounded to the nearest whole per cent) from Australia, the USA, Japan and other countries for 2007.

a What percentage of the total number of visitors comes from other countries?

b What percentage of the total number of visitors comes from Australia, the USA and Japan combined?

Percentage of international visitors to Papua New Guinea in 2007

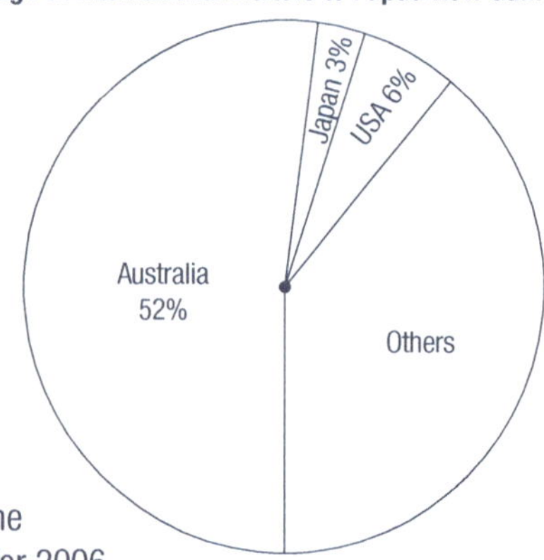

8 Draw a pie chart to represent the percentage of visitor arrivals (rounded to the nearest whole per cent) from Australia, the USA, Japan and other countries for 2006.

9 On the pie chart you have drawn for 2006:

a What percentage of the total number of visitors comes from other countries?

b What percentage of the total number of visitors comes from Australia, the USA and Japan combined?

As discussed previously, visitors to Papua New Guinea fall into different categories depending on the purpose of their visit. Holiday and leisure visitors are generally the most welcomed because they usually stay longer than other visitors and thus contribute more to the local economy.

Of the total number of visitors from Australia, Japan and the USA, this table shows the number who came for a holiday from 2001 to 2007.

Holiday visitors to Papua New Guinea							
Country	**2001**	**2002**	**2003**	**2004**	**2005**	**2006**	**2007**
Australia	5423	5194	6310	7131	7278	10232	13631
Japan	1724	2821	2805	2584	4315	3190	2374
USA	2669	3329	1799	2630	2757	3674	2838

Use the information in the table to help you answer questions 10 and 11.

10 Calculate the percentage increase or decrease in holiday visitors from 2006 to 2007 for each of the three countries. Make sure you indicate if it is an increase or decrease for each country.

11 Calculate the percentage increase in holiday visitors from 2001 to 2007 for:

a Australia b Japan c the USA

d Which country had the largest percentage increase?

e Which country had the smallest percentage increase?

12 Use the table above and the table titled 'Total visitors to Papua New Guinea' to help you calculate:

a the percentage of total Australian visitors who were holiday visitors in 2007

b the percentage of total American visitors who were holiday visitors in 2007

c the percentage of total Japanese visitors who were holiday visitors in 2007

Challenge

Draw a line graph for Australia, Japan or the USA using data from the table above to show the holiday visitors for the years 2001 to 2007.

Help Box

If we want to increase or decrease a number by a certain percentage we can do it in two steps. The first step is to find the amount of increase or decrease and the second step is to add it to or subtract it from the original number.

For example, to increase 360 by 10% we first find 10% of 360 (36) and then we add it to 360 (360 + 36 = 396).

We can also do this in one step by saying $360 \times 1.1 = 396$. The 1 in 1.1 represents the original amount and the 0.1 represents the 10% increase.

To decrease 360 by 10% we say $360 \times 0.9 = 324$. The 0.9 represents the whole amount (1) less the decrease of 10% (0.1).

13 Use one step to decrease these numbers by 10%.

a 90 b 160 c 270

d 340 e 6300

14 Use one step to increase these numbers by 25%.

a 60 b 80 c 124

d 172 e 316

15 In 2008, if each country in the table 'Holiday visitors to Papua New Guinea' had an increase in holiday visitors of 30% from 2007, how many holiday visitors would there be from:

a Australia b Japan c the USA?

Lesson 6 Cost of accommodation around the world

Hotel prices vary greatly around the world, and even within cities. Many hotels change their prices depending on demand so that when there are a lot of tourists the prices increase, and when there are not many tourists prices drop.

Prices per night for a twin room at a selection of hotels around the world are shown below. The prices have been converted to kina to make comparisons easy.

Singapore	K449.88	London (England)	K688.82
Tokyo (Japan)	K625.76	New York (USA)	K895
Bangkok (Thailand)	K330.64	Rome (Italy)	K719.18
Dubai (UAE)	K399.96	Paris (France)	K759.24
Port Moresby (PNG)	K552	Stockholm (Sweden)	K884.38
Nadi (Fiji)	K700.62	Moscow (Russia)	K961.67
Sydney (Australia)	K536.40	Auckland (New Zealand)	K359.44

Use the hotel prices above to help you work out answers to questions 1 to 13.

1 If you stayed for 4 nights at each of the following places, how much would it cost?

a Bangkok b Rome c New York
d Sydney e Auckland

2 If you stayed for 3 nights at each of the following places, how much would it cost?

a Paris b Tokyo c Stockholm
d Moscow e Port Moresby

3 If ten adults were travelling together and booked five twin rooms in London for 2 nights, how much would it cost in total?

4 A tour group of 20 adults booked into ten twin rooms in Singapore. Twelve of the adults booked for 3 nights and the other eight booked for 5 nights. What was the total cost?

5 Find the price difference between a room in Moscow and a room in:

a Dubai b Paris c Rome
d Bangkok e Sydney

6 In which cities are the tourists who are described below?

a Two adults paid a total of K2156.64 for a twin room for 6 nights.

b Two adults paid a total of K2653.14 for a twin room for 3 nights.

c Four adults paid a total of K2503.04 for two twin rooms for 2 nights.

d Six adults paid a total of K8407.44 for three twin rooms for 4 nights.

7 If two adults shared the price of a twin room, how much would they each pay for 1 night at the following hotels?

a Rome b London c Auckland
d Singapore e Dubai

8 If two adults shared the price of a twin room, how much would they each pay for 3 nights at the following hotels?

a Stockholm b Moscow c Nadi

d Paris e New York

9 If hotels in the following cities each increased their prices by 10%, find the new prices using the one-step method outlined in the previous lesson.

a Port Moresby b New York

c Sydney d Bangkok

e Singapore

10 If hotels in the following cities each decreased their prices by 10%, find the new prices using the one-step method outlined in the previous lesson.

a Paris b Auckland c Dubai

d Moscow e Tokyo

11 Some hotels give 5% discount if a reservation is made at least 60 days in advance. If the following hotels were booked 60 days in advance, what would be the cost per night?

a New York b Nadi c Rome

d London e Port Moresby

Most hotels allow three people to share a twin room by providing a roll-away bed. Many hotels charge an extra 35% of the full room price if the third person is an adult and an extra 20% of the full room price if the third person is a child. For example, if the room price is K552, an extra K193.20 (35% of K552) would be added for an adult making a total of K745.20 and an extra K110.40 (20% of K552) would be added for a child making a total of K662.40.

12 Work out the total cost per room for the following situations:

a three adults sharing a twin room in Sydney for 1 night

b two adults and one child sharing a twin room in New York for 1 night

c three adults sharing a twin room in Bangkok for 2 nights

d two adults and one child sharing a twin room in Paris for 3 nights

e three adults sharing a twin room in Stockholm for 3 nights

13 If three adults shared a twin room for 1 night in each of the following cities, how much would each person pay if they shared the total cost equally?

a New York b Nadi c Tokyo

d Dubai e Auckland

This table shows the prices charged by two hotels for a twin room during different seasons.

	Low season	Regular season	High season
Hotel A	K360.37	K402.44	K500.52
Hotel B	K435.26	K595.37	K803.53

Use the prices in the table to answer questions 14 to 17.

14 How much more per room does it cost to stay at:

a Hotel A during high season than during low season?

b Hotel B during high season than during regular season?

c Hotel B during regular season than during low season?

15 If two people share the cost for 3 nights at Hotel B, how much will they each pay during:

a high season b low season

c regular season?

16 If two people share the cost for 5 nights at Hotel A, how much will they each pay during:

a regular season b high season

c low season?

17 Round each room price to the nearest dollar and then calculate the following percentage increases:

a from low season to regular season for Hotel A

b from regular season to high season for Hotel A

c from low season to regular season for Hotel B

d from regular season to high season for Hotel B

Lesson 7 Cargo and passenger statistics

Every year more than one million tonnes of cargo is handled at many of the busiest airports. The figures below show the weight in metric tonnes of the cargo handled at eight different airports for the years 2005 to 2007.

	Total cargo (metric tonnes)		
Airport	**2005**	**2006**	**2007**
Memphis, Tennessee, USA	3 598 500	3 692 081	3 840 574
Hong Kong	3 433 349	3 609 780	3 772 673
Anchorage, Alaska, USA	2 553 937	2 691 395	2 826 499
Frankfurt, Germany	1 962 927	2 127 646	2 169 025
Charles de Gaulle, Paris, France	2 010 361	2 130 724	2 005 160
Singapore	1 854 610	1 931 881	1 918 159
Los Angeles, USA	1 938 430	1 907 497	1 877 876
London, Heathrow, UK	1 389 589	1 343 930	1 395 909

1 Over the 3 years, how much cargo was handled at each of the following airports?

a Singapore b Frankfurt c Memphis d London e Anchorage

2 Calculate the mean weight over the 3 years at each of these airports:

a Charles de Gaulle b Los Angeles c Hong Kong

3 Which airports handled less cargo in:

a 2007 than in 2006 b 2006 than in 2005?

4 Calculate the percentage decrease in the amount of cargo handled at:

a Singapore from 2006 to 2007
b London from 2005 to 2006
c Los Angeles from 2006 to 2007
d Charles de Gaulle from 2006 to 2007

5 Calculate the percentage increase in the amount of cargo handled at:

a Hong Kong from 2006 to 2007
b Frankfurt from 2005 to 2006
c Memphis from 2006 to 2007
d Anchorage from 2005 to 2006

6 If the amount of cargo handled in 2008 increases by 3% from the 2007 figures, how many metric tonnes of cargo (to the nearest tonne) would each of the following airports handle?

a Singapore b Memphis c Anchorage d London

7 If the amount of cargo handled in 2008 decreases by 3% from the 2007 figures, how many metric tonnes of cargo (to the nearest tonne) would each of the following airports handle?

a Charles de Gaulle b Los Angeles c Hong Kong d Frankfurt

Each year millions of passengers pass through the world's busiest airports. The figures below show the total number of passengers from 2005 to 2007 who arrived at, departed from or who were in transit through ten of the world's busiest airports.

Airport	2005	2006	2007
Atlanta, USA	85 907 423	84 846 639	89 379 287
Chicago, USA	76 510 003	77 028 134	76 159 324
London, UK	67 915 403	67 530 197	68 068 554
Tokyo, Japan	63 282 219	65 810 672	66 671 435
Los Angeles, USA	61 489 398	61 041 066	61 895 548
Paris, France	53 798 308	56 849 567	59 919 383
Frankfurt, Germany	52 219 412	52 810 683	54 161 856
Amsterdam, Netherlands	44 163 098	46 065 719	47 793 602
Madrid, Spain	41 940 059	45 501 168	52 122 214
Beijing, China	41 004 008	48 654 770	53 736 923

8 For 2007 calculate how many more passengers passed through Atlanta airport than:

a Tokyo b Beijing c Paris
d Amsterdam e Madrid

9 Calculate the mean number of passengers who passed through each of these airports over the three years:

a Chicago b Los Angeles c Beijing

10 Calculate the percentage increase in the number of passengers who passed through:

a Beijing from 2006 to 2007
b Frankfurt from 2006 to 2007
c Madrid from 2005 to 2006
d Paris from 2005 to 2006

11 Calculate the percentage decrease in the number of passengers who passed through:

a Atlanta from 2005 to 2006
b Chicago from 2006 to 2007

12 The number of passengers passing through Paris airport increased by 5% from 2004 to 2005. Use the figure for 2005 in the table above to calculate the number of passengers who passed through Paris airport in 2004.

13 If the number of passengers in 2008 increases by 20% from the 2007 figures, how many passengers would pass through each of the following airports?

a Los Angeles b Madrid
c Amsterdam d Tokyo

Challenge

Choose three of the airports listed in the table and draw a comparative bar graph that shows the number of passengers passing through each airport from 2005 to 2007. Round each figure to the nearest million to make it easier to draw your graph.

Lesson 8 Some interesting comparisons

Some large countries have quite a small population and some small countries have a large population. Population figures only allow us to compare the number of people living in different countries. If we want to compare population densities, we need to know the land area of countries so we can calculate the number of people per square kilometre.

Help Box

To calculate the number of people per square kilometre (population density) we divide the population by the land area:

$$\frac{\text{Population}}{\text{land area}} = \text{people per square kilometre (population density)}$$

Example: Vanuatu has a population of 208 870 and a land area of 12 190 km^2, so $\frac{208\,870}{12\,190} = 17.13$

This means Vanuatu has 17.13 people per square kilometre.

It is impossible to have 17.13 people, but the two decimal places are sometimes useful when making comparisons between countries with similar population densities.

This table shows the approximate population and land area in square kilometres of 15 countries.

Country	Population	Land area (km^2)
China	1 313 973 710	9 326 410
India	1 095 352 000	2 973 190
USA	298 444 220	9 161 920
Indonesia	245 452 740	1 826 440
Brazil	188 078 230	8 456 510
Pakistan	165 803 560	778 720
Bangladesh	147 365 350	133 910
Australia	20 264 080	7 617 930
Hong Kong	6 940 430	1 042
Papua New Guinea	5 670 540	452 860
Singapore	4 492 150	689
New Zealand	4 076 140	268 020
Vanuatu	208 870	12 190
Tonga	114 690	718
Monaco	32 540	1.95

1 Use the data in the table to calculate the population density of each country.

2 List each country in order according to their population density.

3 Name the country that has:

a the largest population

b the largest land area

c the largest population density

4 Name the country that has:

a the smallest population

b the smallest land area

c the smallest population density

5 Write sentences to explain why:

a Monaco has the most people per square kilometre of the countries listed.

b Australia has the least number of people per square kilometre of the countries listed.

6 List the countries that have:

a fewer than 50 people per square kilometre

b more than 5000 people per square kilometre

7 If the population of the following countries increased by 100 000, how many people per square kilometre would there be in:

a Australia

b Papua New Guinea

c New Zealand

d Tonga?

Challenge

Calculate the population density of your classroom per square metre.

Help Box

If we know the population density of a country and its land area, we can calculate its population by multiplying population density by land area:

$$\text{population} = \text{population density} \times \text{land area}$$

If we know the population density of a country and its population, we can calculate its land area by dividing its population by its population density:

$$\text{land area} = \frac{\text{population}}{\text{population density}}$$

8 Calculate the population of countries with the following population density and land area.

a 49.59 people per km^2 and 18 270 km^2

b 0.14 people per km^2 and 410 450 km^2

c 141.45 people per km^2 and 468 km^2

d 12.83 people per km^2 and 18 580 km^2

9 Calculate the land area of countries with the following population and population density.

a 2 554 000 people and 1.63 people per km^2

b 13 956 980 people and 11 people per km^2

c 11 668 460 people and 15.7 people per km^2

d 20 100 people and 43.7 people per km^2

Education is another topic that allows us to make interesting comparisons between countries. The table in the next column shows the years of formal schooling on average that students receive in twenty different countries. The countries listed in the column on the left are the top ten in the world. The remaining ten countries, although in order, are selected from the many that were available.

Country	Years	Country	Years
USA	12	Japan	9.5
Norway	11.8	Fiji	8.3
New Zealand	11.7	France	7.9
Canada	11.6	Singapore	7
Sweden	11.4	Malaysia	6.8
Australia	10.9	Thailand	6.5
Switzerland	10.5	China	6.4
Germany	10.2	Indonesia	5
Finland	10	Papua New Guinea	2.9
Poland	9.8	Nepal	2.4

10 How many more years of schooling on average do people in Canada receive than people in:

a Australia b Malaysia c Poland

d China e Papua New Guinea?

11 How many more years of schooling do people in Nepal need to reach the average in:

a Japan b Australia c Indonesia

d Fiji e Thailand?

12 How many more years of schooling do people in Papua New Guinea need to reach the average in:

a France b China c Singapore

d Malaysia e Poland?

13 Use the same scale to draw each graph described below.

a Draw a bar graph to display the average years of schooling of people who live in the ten countries listed in the left column of the table.

b Draw a bar graph to display the average years of schooling of people who live in the ten countries listed in the right column of the table.

14 Write the number of countries that have 10 or more years of schooling on average as:

a a fraction of all countries listed above

b a percentage of all countries listed above

15 Write the number of countries that have less than 7 years of schooling on average as:

a a fraction of all countries listed above

b a percentage of all countries listed above

Learning Unit  Additional Learning, Revision and Assessment

Strand: Chance and Data

Statistics	Outcome 8.4.1	Interpret information presented statistically

Lesson 1: Additional Learning

Focus: Using angles to construct and interpret pie charts.

Lesson 2: Revision

Mapping conventions
Rates, percentage and decimal problems
Rotational symmetry
Temperature and time calculations
Statistics

Lesson 3: Assessment Task 1

Practical investigation—group work
Assessment Task 1 will assess learning outcomes 8.1.2, 8.1.6, 8.2.14, 8.2.16 and 8.3.4.

40 marks

Lesson 4: Assessment Task 2

Test of basic skills and routine applications
Assessment Task 2 will cover the learning outcomes from across the topic 'The Tourism Market'.
The test will assess the extent to which students can:

- Demonstrate an understanding of the mathematical concepts
- Correctly choose and apply mathematical techniques to solve problems

60 marks

Total 100 marks

Lesson 1 Additional learning

Using angles to interpret and construct pie charts

In Grade 7 you learned how to create pie charts or pie graphs by using degrees and estimating angles. This lesson will revise and extend that learning.

Look at the circle and answer the following questions.

1 What is the size of:

- a Angle A
- b Angle B
- c Angle C
- d Angle D?
- e How many degrees are in a circle?

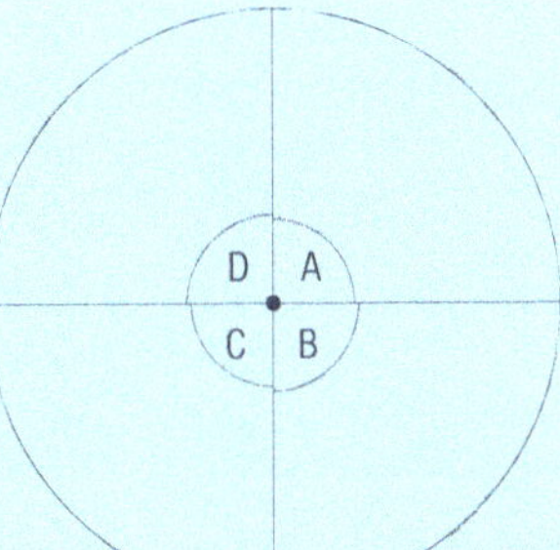

2 Look at the pie chart shown here.

- a Copy the table below into your workbook and complete the first column by estimating the size of each angle in the pie chart in degrees.

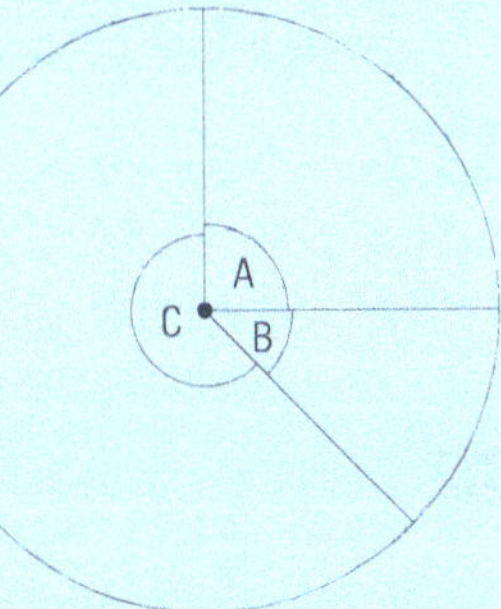

Angle	Estimate of angle size	Number of degrees	Fraction (out of 360)	Fraction (simplest form)
A				
B				
C				

- b Use a protractor to measure each angle in the pie chart and write the measurements in the second column of the chart.
- c In the third column write each number of degrees as a fraction of 360.
- d In the last column write each fraction in its simplest form.

3 The pie chart in question 2 represents a survey of 64 students to find their favourite sport out of soccer, volleyball or touch rugby. The results showed that touch rugby was the most popular, followed by volleyball and then soccer.

- a How many students preferred touch rugby?
- b How many students preferred volleyball?
- c How many students preferred soccer?

4 Draw three tables like the one below in your workbook.

Angle	Estimate of angle size	Number of degrees	Fraction (out of 360)	Fraction (simplest form)
A				
B				
C				
D				

Complete each table by referring to the pie charts shown here.

a

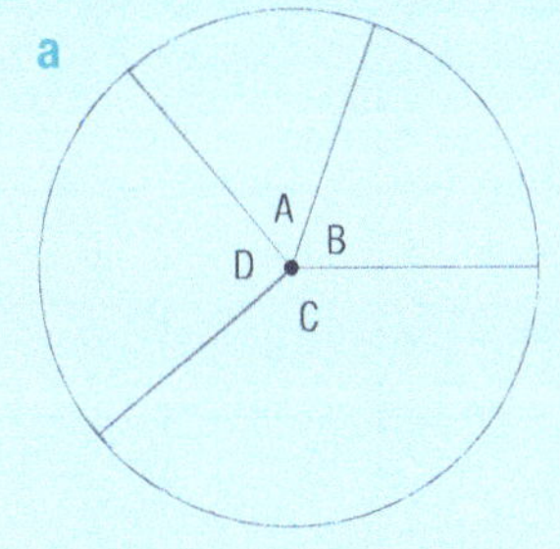

b

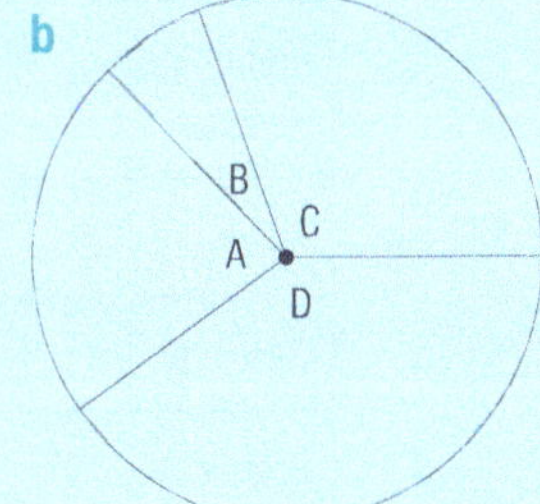

c

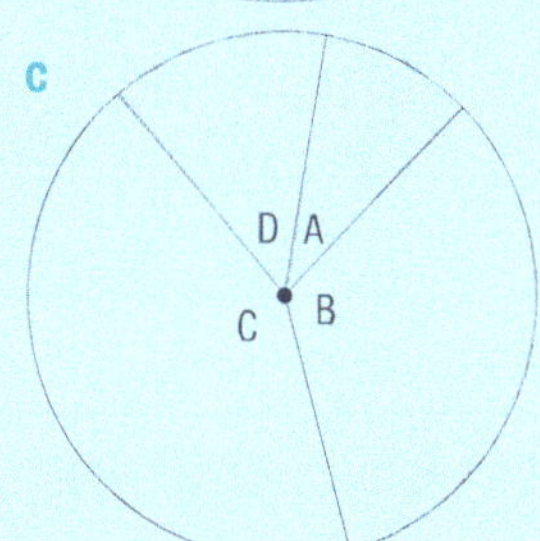

5 If the pie chart in question 4a represents an amount of K144 to be split between four people (A, B, C and D), how much does each person get?

6 If the pie chart in question 4c represents a survey of 2160 people to find their favourite colour out of red, blue, green and yellow, and the results showed that blue was the most popular, followed by red, yellow and then green, how many people chose:

a blue b red

c yellow d green?

7 This pie chart represents the way Mr Wealthy wants his K60 000 to be shared when he dies. Use a protractor to find the angle, then work out the fraction and the value of each share.

	Beneficiary	Angle (in degrees)	Fraction	Value of share
a	Children's Hospital			
b	Son			
c	Daughter			
d	Grandson			
e	Grand-daughter			
f	Brother			
g	Best friend			

Help Box

This table shows the results of a survey carried out with 60 teenagers.

Volleyball	17
Athletics	12
Soccer	24
Basketball	7

To display this data on a pie chart we divide 360° by the number of teenagers:

$$360° \div 60 = 6°,$$

so 6° will represent one teenager.

For example, to represent 17 teenagers playing volleyball:

$$17 \times 6° = 102°,$$

so an angle of 102° can be drawn on the pie chart.

8 If a pie chart is drawn to display the data in the table in the Help Box, calculate the number of degrees that will represent the other three sections of the pie chart.

9 a How many degrees would represent each person if 90 teenagers were surveyed?

b How many degrees would represent each person if 120 teenagers were surveyed?

10 Draw a pie chart to represent the data about the 60 teenagers in the Help Box above. Use a pair of compasses to draw a circle and a protractor to divide it into sections according to the angle sizes you have calculated. Label each section with the name of the sport it represents and give your graph a title.

11 The results of a survey about most liked vegetables are shown below.

a How many people were surveyed?

b If the data is displayed on a pie chart, how many degrees will represent one person?

c Use a pair of compasses and a protractor to draw a pie chart to represent the vegetable survey data. Label each section with the name of the vegetable it represents and give your graph a title.

Kaukau	9
Beans	11
Corn	6
Sweet potato	7
Taro	7

Lesson 2 Revision

Check your understanding of this topic using the following exercises.

Unit one

1 A map has a scale of 1:250 000. What lengths on the map represent these distances on the ground?

a 5 km b 20 km c 7.5 km d 1.25 km e 15 km

2 A map has a scale of 1:500 000. What actual distances on the ground are represented by these measurements on the map?

a 2 cm b 8 cm c 4.5 cm d 5.2 cm e 9.6 cm

3 Convert the second measurement in each scale to centimetres, then write the scale as a ratio.

a 1 cm = 2 km b 1 cm = 5 m c 1 cm = 8.5 km d 1 cm = 100 m

4 Calculate the distance travelled at the following speeds and times.

a 200 km per hour in 45 minutes
b 450 km per hour in $1\frac{1}{2}$ hours
c 320 km per hour in $1\frac{3}{4}$ hours
d 240 km per hour in 1 hour 20 minutes

5 Calculate the flying time of planes that travel:

a 1000 km at 250 km per hour
b 1250 km at 500 km per hour
c 920 km at 200 km per hour
d 1360 km at 320 km per hour

6 Add 10% to each of the following money amounts.

a K259 b K580 c K327 d K456 e K114 f K295

7 Calculate the discount on the following money amounts.

a 50% of K236 b 20% of K318 c 40% of K260 d 25% of K438
e 15% of K184 f 30% of K354 g 20% of K192 h 15% of K516

8 a Which of the following shapes have rotational symmetry?

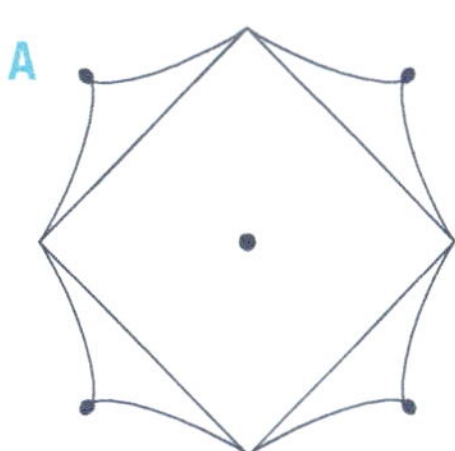

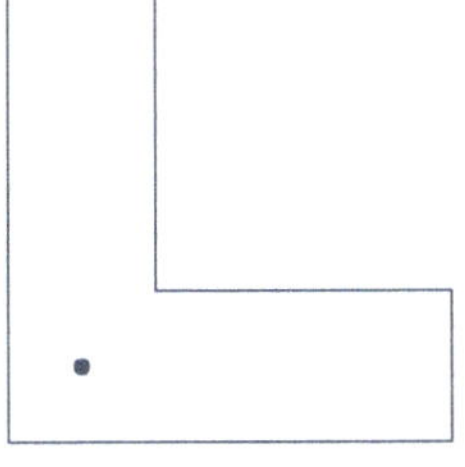

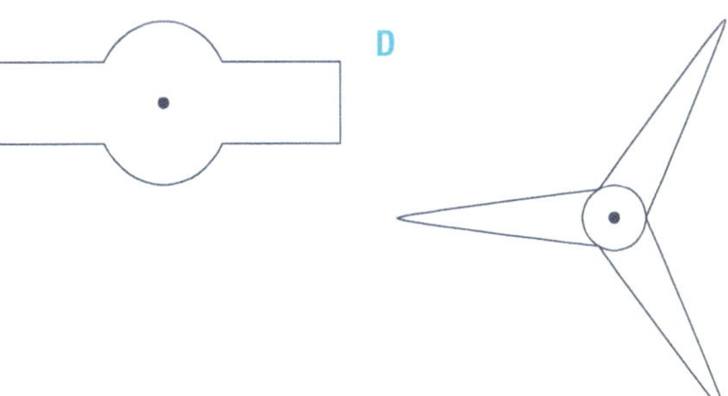

b Which shape has rotational symmetry of order 2?

9 If a car's average fuel consumption is 7.5 litres per 100 km, calculate, to the nearest whole number, the fuel consumption for the following journeys.

a 510 km b 350 km c 424 km d 272 km e 88 km f 156 km

10 Calculate the fuel consumption per 100 kilometres of a car that travels:

a 320 km and uses 25 litres of fuel
b 215 km and uses 18 litres of fuel
c 260 km and uses 16 litres of fuel
d 435 km and uses 28 litres of fuel

11 78 265 people visited an island in one year. Of those people, 32 018 were children.

a Find the percentage of the people that were children. Write your answer to the nearest whole number.
b 19 874 of the children were girls. What percentage of the children is this? Write your answer to the nearest whole number.

Unit two

12 Complete these sentences by entering the correct direction. Use an atlas to help you.

a Port Moresby is _____ of the equator.
b London is _____ of the equator.
c Tokyo is _____ of the Tropic of Cancer.
d Sydney is ___ of the Tropic of Capricorn.
e Bangkok is _____ of the Tropic of Cancer.
f Cairo is _____ of the prime meridian.
g New York is _____ of the prime meridian.
h Perth is _____ of the prime meridian.

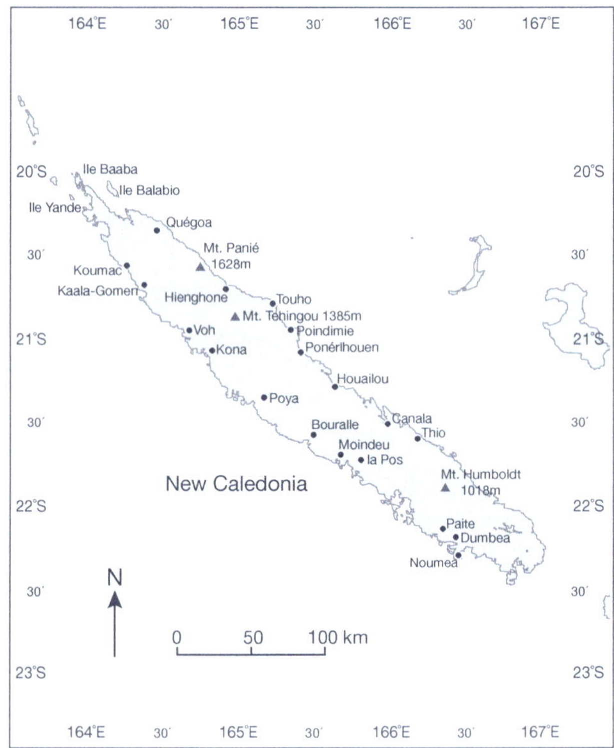

13 Locate the place on the map of New Caledonia that is at each of the following latitudes and longitudes.

a 22°15′S, 166°30′E
b 20°45′S, 165°15′E
c 20°35′S, 164°45′E
d 21°20′S, 165°10′E

14 Give the location of these places using both degrees and minutes.

a Thio
b Koumac
c Voh
d Mt Humboldt

15 When it is 5 p.m. in Port Moresby, it is 2 p.m. in Jakarta.

a If a plane departs Port Moresby at 5 p.m. local time and lands in Jakarta $5\frac{1}{2}$ hours later, what will the local time be in Jakarta?
b If another plane departs Jakarta at 8 a.m. local time and lands in Port Moresby $5\frac{1}{2}$ hours later, what will the local time be in Port Moresby?

16 When it is 11 a.m. in Sydney, it is 1 p.m. in Auckland.

a If a plane departs Auckland at 1 p.m. local time and lands in Sydney 3 hours later, what will the local time be in Sydney?
b If another plane departs Sydney at 10.30 a.m. local time and lands in Auckland 3 hours later, what will the local time be in Auckland?

17 Calculate the approximate fuel consumption per 100 kilometres of:

a a plane with a fuel capacity of 33 500 L and a range of 4750 km

b a plane with a fuel capacity of 29 250 L and a range of 4200 km

18 Calculate the approximate flying time for the following planes if their tanks are full:

a a plane with a range of 13 200 km and a cruising speed of 825 km per hour

b a plane with a range of 5250 km and a cruising speed of 700 km per hour

19 Two hundred and seventeen passengers each checked in baggage before their flight. The mean weight of each bag was 18.45 kg.

a What is the total mass of the 217 bags?

b How much less than 4500 kg is this?

20 The total mass of baggage checked in for 193 passengers was 3734.55 kg. What is the average mass of each passenger's bag?

21 If an airline charges K27.20 per kilogram for excess baggage, how much will the following excess baggage cost?

a 8.5 kg b 13.2 kg c 5.8 kg d 22.3 kg e 18.65 kg

22 If K1 buys 0.40 Australian dollars, how much in Australian dollars would you get for these amounts?

a K200 b K50 c K1000 d K425 e K675

23 Sari exchanged euros at a rate of K1 = €0.26. How many kina would he get for these amounts?

a €250 b €180 c €300 d €545 e €485

24 Calculate the price when 25% is taken off these full fares.

a K6150 b K8066 c K6995 d K9127 e K5894

25 The following fares represent 80% of the full adult fare. Calculate each full adult fare.

a K5720 b K7440 c K3216 d K4672 e K6392

Unit three

26 Convert these measurements in metres to feet by using the rate 1 metre = 3.28 feet. Round your answers to the nearest tenth.

a 30 m b 65 m c 48 m d 94 m e 73 m f 109 m

27 Use the conversion rate 1 foot ≈ 0.3 metre to convert these lengths to metres.

a 85 feet b 175 feet c 238 feet d 124 feet e 315 feet f 266 feet

28 Obtain a more accurate answer by converting the lengths in question 27 to metres by dividing by 3.28. Round your answers to the nearest tenth.

29 Convert these measurements in kilometres to miles by using the conversion rate 1 km = 0.62 miles. Round your answers to the nearest tenth.

a 67 km b 127 km c 88 km d 216 km e 181 km f 245 km

30 Use the conversion rate 1 mile ≈ 1.6 km to convert these lengths to kilometres.

a 95 miles b 123 miles c 72 miles d 158 miles e 230 miles f 197 miles

31 Convert the distances in question 30 to kilometres by dividing by 0.62. Round your answers to the nearest tenth.

32 Calculate the percentage increase from:

a 54 500 to 62 700 b 37 890 to 48 600 c 82 325 to 91 650 d 68 170 to 79 400

33 Calculate the percentage decrease from:

a 85 750 to 69 800 b 102 500 to 96 850 c 71 008 to 67 440 d 92 065 to 85 220

34 Increase these numbers by 15%.

a 60 b 75 c 140 d 96 e 108 f 220

35 Decrease these numbers by 5%.

a 80 b 106 c 260 d 154 e 215 f 340

36 A hotel room costs K419.56 per night. What would it cost to stay for the following numbers of nights?

a 3 nights b 7 nights c 5 nights d 2 nights

37 If three people shared the costs of the following hotel rooms equally, how much would each person pay?

a K754.50 b K615.30 c K688.50 d K829.20 e K726.90 f K570.60

38 Calculate the number of people per square kilometre for a country that has a population of 815 450 and a land area of 12 500 square kilometres. Round your answer to two decimal places.

39 Calculate the population of a country that has a population density of 47.6 people per square kilometre and a land area of 96 085 square kilometres.

40 Calculate the land area of a country that has 5 600 000 people and a population density of 87.5 people per square kilometre.

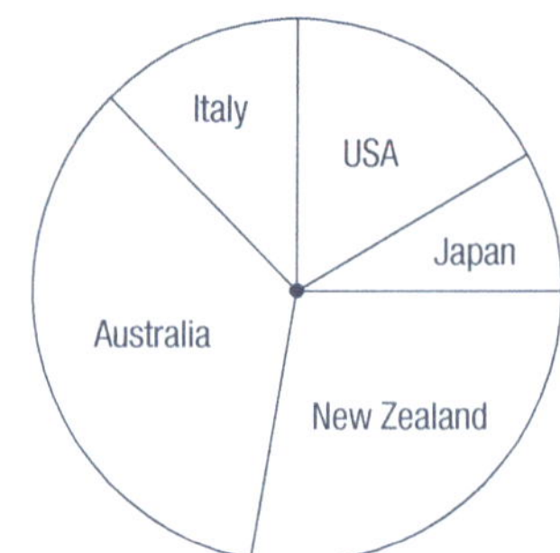

Additional learning unit

41 A travel agency constructed this pie chart to show the holiday destinations of 1800 of its clients. Measure each angle with a protractor and calculate the number of people who travelled to each country.

Assessment

Investigation

This assessment task gives you the chance to show what you understand about mapping, rates, decimals and the relationships between locations, time and temperature in a real-life setting. The work you have done in the topic 'The Tourism Market' and especially in the unit 'Travelling Overseas' will help you to complete this task.

Task 1: Practical investigation – group work

Imagine that your group has just been told that it has won a competition to visit another country. Work together as a group to choose a destination and find the following information about that destination.

- Look at a world map in an atlas and decide on one country that you would like to visit. Name a major city in that country that would be your destination, and use latitude and longitude to give its approximate position.
- Calculate the approximate distance of this major city from Port Moresby.
- Choose a suitable aircraft from the list in Lesson 6 in Learning Unit 2 (Travelling Overseas) and calculate the flying time based on the distance of your chosen destination from Port Moresby and the given cruise speed of the aircraft. Show your calculations.
- Use the World Time Zones map in Lesson 5 in Learning Unit 2 to help you calculate the time difference between Port Moresby and your destination.
- List three possible departure times (morning, afternoon and evening) and calculate the equivalent local times at your destination.
- Make a table that shows your three departure times, your flying time and the local time when you arrive at your destination.
- Use what you have learned about location and temperature to estimate the daytime temperature of your destination. Give reasons to support your estimate.
- Find out which currency is used in your chosen country and research the current exchange rate. Make a table that shows various amounts in kina and their equivalent amounts in the currency of your chosen country. Show evidence of your calculations.
- Make a group decision about the most effective way of presenting your completed work and then do this.

  Your group will be assessed on your:
  - ability to work productively as a group
  - evidence of reasonable estimations made prior to calculations
  - choice of appropriate problem-solving strategies
  - correctness of calculations
  - presentation of tables associated with the task
  - overall effort and persistence
  - general presentation of final product
  - ability to complete the task in the time allocated.

Total 40 marks

Assessment

Test

Task 2: Test of basic skills and routine applications

Fractions

1 How much smaller are maps with the following scales than the actual places they represent? Write each answer as a fraction in its simplest form.

a 1 cm = 50 m b 1 cm = 2 km c 5 cm = 2.5 km d 2 cm = 0.5 km (4 marks)

2 Calculate:

a $\frac{2}{5}$ of K260 b $\frac{2}{3}$ of K726 c $\frac{3}{4}$ of K180.60 d $\frac{3}{8}$ of K452 (2 marks)

Ratios and Rates

3 How much VAT (Value Added Tax) in the ratio 1 : 10 would be added to these amounts?

a K150 b K300 c K87 d K245 (2 marks)

4 A plane is flying at a speed of 680 km per hour. How far has it flown after $2\frac{3}{4}$ hours? (1 mark)

5 How long will it take an aircraft to fly 4750 km at a rate of 500 km per hour? (1 mark)

6 Convert the second measurement in each scale to centimetres and then write the scale as a ratio.

a 1 cm = 10 km b 1 cm = 750 m c 1 cm = 45 km d 1 cm = 8.5 km (4 marks)

Decimals

7 If a car uses 7.3 L of petrol per 100 km, how much will be used for the following journeys?

a 250 km b 175 km c 84 km d 132 km (2 marks)

8 If petrol costs K2.94 per litre, how much does it cost for the following amounts?

a 25 litres b 30 litres c 18 litres d 44 litres (2 marks)

9 A hotel room for 3 people cost K516.60 per night. If three people shared the cost equally, how much did each person pay? (1 mark)

10 If an airline charges K32.40 per kilogram for excess baggage, how much will the following excess baggage cost?

a 8 kg b 12.5 kg c 6.25 kg d 17.8 kg (4 marks)

11 A passenger pays a total of K412.50 for excess baggage. If the rate is K27.50 per kilogram, how many kilograms of extra baggage does the passenger have? (2 marks)

12 If K1 buys 0.51 New Zealand dollars, how much in New Zealand dollars would you get for these amounts?

a K100 b K250 c K180 d K600 (2 marks)

13 If K1 buys 2.80 South African rand (R), how many kina would you get for each of these amounts?

a R140 b R336 c R700 d R105 (2 marks)

Percentages

14 What is the total cost of each hotel room after 10% VAT is added?

a K430 b K375 c K524 d K293 (2 marks)

Assessment

Test

15 Write these numbers as percentages.

a 2615 out of 6950 b 15 220 out of 21 365 c 18 965 out of 24 140 (3 marks)

16 Calculate the percentage increase or decrease from:

a 12 895 to 9640 b 56 500 to 78 950 c 41 250 to 89 800 d 74 320 to 58 625 (4 marks)

17 Increase these numbers by 5%.

a 90 b 210 c 135 d 172 (2 marks)

18 Decrease these numbers by 20%.

a 120 b 86 c 278 d 164 (2 marks)

19 These prices represent 85% of the full price. Calculate each full price.

a K765 b K2975 c K5270 d K3825 (4 marks)

Direction

20 A map has a scale of 1:15 000. What lengths on the map represent these distances on the ground?

a 300 m b 1.5 km c 0.75 km d 975 m (2 marks)

21 Using the scale 1:5000 draw a representation of a square with sides of 150 metres. (2 marks)

Maps and Coordinates

22 A location was given as 40°N, 85°W. In your own words explain what this means. (2 marks)

Weight

23 The mean weight of 216 bags is 17.85 kg.

a What is the total weight of all the bags? b How much less than 4500 kg is this? (2 marks)

24 The mean weight of five bags was 20.0 kg. What might each bag have weighed? (2 marks)

Time

25 When it is 6 p.m. in Port Moresby, it is 10 a.m. in Cape Town.

a If a plane departs Port Moresby at 6 p.m. local time and lands in Cape Town 18 hours later, what will the local time be in Cape Town?

b If another plane departs Cape Town at 10 a.m. and lands in Port Moresby 18 hours later, what will the local time be in Port Moresby? (2 marks)

26 Tokyo is 3 hours behind Auckland. It takes 11 hours to fly between Tokyo and Auckland.

a If a plane leaves Tokyo at 7 p.m., what will be the local time in Auckland when it lands?

b If a plane leaves Auckland at 11 p.m., what will be the local time in Tokyo when it lands? (2 marks)

Total 60 marks

For Teachers

For Teachers

The Oxford Grade 8 Mathematics Program was developed to reflect the teaching approaches described in the Upper Primary Teachers Guide (2003) produced by the Department of Education, Papua New Guinea. The Teachers Guide provides valuable information about curriculum development, appropriate use of context, selection of suitable learning and teaching strategies and how to plan and carry out effective assessment. The Grade 8 Mathematics Program should be used in conjunction with the Teachers Guide to help teachers implement the Grade 8 Mathematics Syllabus.

The Grade 8 Program comprises four topics across two books. Book A and Book B each contain two topics and it is expected that each topic will take one term to complete.

Each topic is divided into three Learning Units and one Additional Learning, Revision and Assessment unit.

All learning outcomes for each Mathematical Strand have been covered in the Oxford Grade 8 Mathematics Program. On later pages a chart matches the learning outcomes to the various Learning Units.

A key focus of the Grade 8 Mathematics books is that they are based on particular contexts. Four broad topics of general interest have been chosen and each Learning Unit examines a specific context within these broader topics.

The mathematical concepts have been organised to build upon each other as students work through the two books. This means that in order to make learning as effective as possible, students should understand the concepts contained in Book A before beginning Book B. However, depending on your students' needs, you may decide to alter the order in which they complete the topics.

Planning your teaching

In Upper Primary 180 minutes per week has been allocated to mathematics. The National Department of Education's policy is to allow flexibility in timetabling and you may, for example, plan to teach five 36-minute lessons or four 45-minute lessons a week.

As a general guide teachers should aim to finish one topic in the Grade 8 Mathematics book each term. The individual lessons vary in length but each would be expected to take a minimum of 36 minutes to complete and in some cases may take two or three lessons. Plan to spend about 3 weeks on each Learning Unit and one week on the Revision Unit which includes a formal test.

Using contexts

Recent reforms in education have highlighted the need to support students so they can use the mathematics they learn in realistic, practical situations. Using a context for teaching mathematics has the added advantage of helping students to see the meaning of the concepts and of making the posing of mathematical problems easier to understand.

It is important that the curriculum builds links for students between the mathematics that they do at school and its application to real-life situations.

Explain why you are using the contexts

Linking new mathematical concepts to familiar and interesting contexts enables students to integrate their learning in a way that relates more readily to their own lives. The contextual approach requires students to use knowledge and skills from a range of sources to find solutions using the methods and levels of accuracy appropriate to the real-life context.

Discuss the context with students

The context is useful for showing students how the mathematics is used in real-life. Generally, contextualising the mathematics also makes it easier to understand. However, in some cases the context can make the mathematics more difficult to learn. For example, if you want to talk about sea-shells, students from Highlands regions may have difficulty understanding what you mean. It is very important that students understand the context that is being used. Teachers should adapt the language and/or the context to best meet the needs of their students.

Transfer the concept to another context

Mathematical knowledge and skills gained in one context are only useful if they can be applied in some other context. While it is helpful to initially teach a concept using a familiar context, it is important that the student finally be asked to solve a similar problem using a different context. Only then can teachers know that the concept has been learned.

Topic introduction

Each topic in the Grade 8 Mathematics Program begins with a double page spread that provides an overview of the broader context and its relevance to the student. The links between the context and the mathematical concepts for each Learning Unit also appear on this double page spread. Use the topic introduction pages to initiate discussion about the context and to find out what students already know about the real-life situations described and the mathematics involved.

The first lesson in each Learning Unit introduces the specific context of that unit within the broader topic. The introduction brings the context into the realm of the individual student and explores their existing knowledge of the context at a personal level.

Catering for diversity

It is the teacher's responsibility to ensure that all students progress in their learning. The lessons in these books have deliberately used contexts that are sufficiently broad enough to cater for a wide range of student interests and backgrounds.

The books include a Glossary and Help Boxes to scaffold student learning, while the Challenges promote and extend thinking. Similarly the Revision Units provide an opportunity for students to demonstrate their understanding by undertaking a student-centred investigation and completing a formal test.

Teaching and learning strategies

Some of the concepts in the books are quite complex and may require a different teaching style. The Papua New Guinea Mathematics Syllabus (2003) promotes the importance of giving students opportunities in mathematics to work cooperatively, discuss and listen to each other's opinions. It is important that students are clear about the problems they are to solve and that they are provided with adequate time to find solutions themselves without being told how to do them. In some cases it may be appropriate for the class to work together on one question rather than work independently on a series of similar questions.

Language

English is the main language of instruction for Upper Primary, although local vernacular can be used to facilitate understanding and reinforce meaning. The Glossary at the back of each book identifies the key words that students need to learn while working through that book. The first few times students meet these key words in the books, teachers should:

- say the word with the class a number of times
- write the word on the board or on a class list or piece of cardboard
- explain the word using real objects, actions or pictures
- demonstrate how to use the word in a simple mathematical sentence
- ask the students to use the word in a simple mathematical sentence
- tell the students to enter the word in their language vocabulary book or the class dictionary.

Materials

The Grade 8 Mathematics books use a range of different materials. Students learn best when they use real materials to help them explore new concepts. Many of the materials that you will need for the program can easily be collected from the students' environment or made from simple items commonly available. Before students use new materials they should be given time to explore them. This will allow them to concentrate on the task when it is finally set instead of wanting to play with the new materials.

In Grade 8 the intention is for students to move progressively away from using real materials when working with numbers to deal directly with symbols. However, with concepts such as fractions, decimals and algebra it may still be preferable to use concrete materials to assist student understanding.

A list of the main materials needed for each topic appears on the double page spread at the start of each topic. It is important that teachers check this list before beginning the topic and ensure that adequate quantities of the materials are available for the students to use. Some of the mathematical equipment listed may need to be purchased or borrowed from other institutions within the community if the school's resources are insufficient.

Materials and equipment should be stored in strong boxes with lids. Students should be told that they are expected to look after materials and to return them to the appropriate box when they have finished using them. Teachers should check all materials regularly and repair or replace any that are damaged.

Assessment

Assessment is the process of finding out what students know. It is important to assess students in order to evaluate what mathematical concepts they understand, how they learn best, what concepts they are ready to learn and how to most appropriately adjust teaching to further support student learning.

Assessment information should be gathered throughout the year using a variety of assessment methods. Written tests and examinations should only form part of the assessment process. Other valuable information can be gathered during everyday teaching by listening to what students say, observing what they do and looking at samples of student work. It is better to collect a small amount of information at many different times throughout the year than to collect a lot of information a few times during the year.

How to evaluate your students

Assessment should be an ongoing process that begins from the first day at school and continues throughout the year. It is important that teachers do not expect students to have mastered all concepts before they have had the necessary experiences and are genuinely able to demonstrate their use of the concepts successfully.

Using open-ended tasks can provide teachers with meaningful information about students that may not be apparent from student responses to closed tasks. The following example shows the difference between an open and closed task about money.

Closed task: What change should be given if K10 is paid for goods that cost K7.35?

Open-ended task: Describe three different ways to give a customer change from K10 for goods that cost K7.35.

There are many possible answers to the open-ended question. A teacher can learn more about what a student understands about money from their responses to the open-ended question than from their one answer to the closed question. Student responses to open-ended questions can show whether they can:

- look at the problem in different ways
- find more than one possible answer
- explain their strategies
- see patterns and make links between their answers

How to record assessment information

Given the variety of assessment strategies that teachers are encouraged to use, it helps if teachers develop a systematic way for recording information about their students. It is important to keep records up to date and ensure that some data is gathered from each student regularly enough to show changes in their skills and knowledge. Assessment should take place at least fortnightly if not weekly.

Some teachers keep student work sample folders that show evidence of the most recent level of student understanding and achievement. Ideally students should make suggestions and discuss with their teacher about which work samples to include in their folder. As a new piece of work shows improved understanding, the older work sample should be removed from the file. Student work sample folders that are kept up to date are a powerful method of monitoring and assessing student progress.

Another strategy is for teachers to keep notes on a class list. Teachers could choose to observe one to two students in each lesson and make comments about what they observe regarding the students' skills, understandings and motivation. The notes could include both positive or negative information and both typical and unusual events. An example of an annotated class list appears below.

Date	**Name**	**Comments**
24 March	Elsie	Added all decimals correctly. Used phrase 'zero point eighty-seven'. (May suggest misconception of decimals as whole numbers instead of parts of whole.)
26 March	Fabian	Confused about placement of decimal point when multiplying decimals by decimals.
26 March	Leti	Completed all decimal additions correctly and asked to show others how to do it.

Tests

These may be short answer or longer exercises. An end of unit test by itself provides information too late to be of value in the classroom. Use the formal test provided in the Revision Unit in conjunction with your on-going forms of assessment. Regular short tests during a unit of work provide timely information about student progress and offer a chance for the teacher to change their teaching approach if required in order to improve student outcomes.

Using assessment information

All assessment information should be used primarily to evaluate student performance as a means of refining the teaching approach. There is no value in collecting assessment information if it is not used to inform teaching for the purpose of improving student performance. If the assessment process highlights gaps in a student's knowledge and understandings, it is a signal for the teacher to try a different approach. Assessment helps the teacher to provide better learning opportunities for the student to ensure the desired outcomes are achieved.

Outcomes Map

Strand	Learning Outcome	Topic	Unit
Number and Application	8.1.1	Gadgets And the Winner is… A Lot Like Me The Tourism Market	In the Tool Shed In the Bag Different People, Different Sizes Growing and Changing Where Do Tourists Go?
	8.1.2	And the Winner is… A Lot Like Me The Tourism Market	Water Sports Growing and Changing Where Do Tourists Go? Travelling Overseas Tourism Comparisons
	8.1.3	Gadgets The Tourism Market	In the Tool Shed Where Do Tourists Go? Tourism Comparisons
	8.1.4	A Lot Like Me The Tourism Market	Growing and Changing Where Do Tourists Go? Travelling Overseas Tourism Comparisons
	8.1.5	Gadgets And the Winner is… A Lot Like Me The Tourism Market	In the Tool Shed The Games People Play Different People, Different Sizes Where Do Tourists Go?
	8.1.6	Gadgets The Tourism Market	Mobile Phones Where Do Tourists Go? Travelling Overseas
	8.1.7	Gadgets	Mobile Phones Additional Learning Unit
	8.1.8	And the Winner is… A Lot Like Me	Additional Learning Unit Additional Learning Unit
Space and Shape	8.2.5	Gadgets	Wheels
	8.2.6	And the Winner is…	In the Bag
	8.2.7	And the Winner is…	Water Sports
	8.2.8	And the Winner is… The Tourism Market	Water Sports Tourism Comparisons
	8.2.9	Gadgets	Wheels
	8.2.10	The Tourism Market	Where Do Tourists Go?
	8.2.11	Gadgets	In the Tool Shed
	8.2.12	Gadgets	In the Tool Shed
	8.2.13	And the Winner is…	In the Bag

Strand	Learning Outcome	Topic	Unit
Space and Shape	8.2.14	The Tourism Market	Where Do Tourists Go? Travelling Overseas
	8.2.16	The Tourism Market	Travelling Overseas
Measurement	8.3.1	A Lot Like Me The Tourism Market	Different People, Different Sizes Travelling Overseas
	8.3.2	A Lot Like Me	Different People, Different Sizes
	8.3.3	A Lot Like Me The Tourism Market	Different People, Different Sizes Travelling Overseas
	8.3.4	Gadgets The Tourism Market	Mobile Phones Travelling Overseas
	8.3.5	Gadgets The Tourism Market	Mobile Phones Where Do Tourists Go? Travelling Overseas
Chance and data	8.4.1	And the Winner is… A Lot Like Me The Tourism Market	Water Sports The Past and the Future Travelling Overseas Tourism Comparisons Additional Learning Unit
	8.4.2	Gadgets A Lot Like Me	Wheels The Past and the Future
	8.4.3	And the Winner is…	The Games People Play
	8.4.4	And the Winner is…	Water Sports
	8.4.5	A Lot Like Me The Tourism Market	Different People, Different Sizes Travelling Overseas Tourism Comparisons
	8.4.6	Gadgets A Lot Like Me	Wheels Different People, Different Sizes
	8.4.7	Gadgets And the Winner is… A Lot Like Me The Tourism Market	In the Tool Shed The Games People Play The Past and the Future Where Do Tourists Go? Travelling Overseas Tourism Comparisons
Patterns and Algebra	8.5.1	And the Winner is…	In the Bag
	8.5.2	And the Winner is… A Lot Like Me	The Games People Play The Past and the Future
	8.5.3	Gadgets A Lot Like Me	Mobile Phones Different People, Different Sizes

Glossary

Glossary

adjacent Next to each other. For example, in this triangle side AB is adjacent to side AC because they have a common vertex (A).

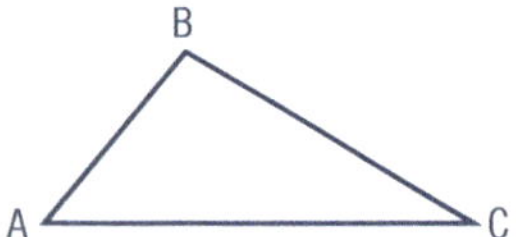

annual Occurring in a year. For example, annual visitors are the number of people who visit something in one year.

anticlockwise It is the opposite direction to clockwise. It is the opposite direction to the movement of the hands on an analogue clock.

approximate Near but not exact.

approximation An answer that is near but not exact. The symbol $\approx$ stands for 'is approximately equal to', for example, $0.59 \approx 0.6$.

area A measure of the total surface of a shape or object. Area is measured in square centimetres (cm^2), square metres (m^2), hectares (ha) or square kilometres (km^2).

attribute A characteristic or property of an object or item. Colour, size, and shape are attributes.

average Also called the *mean* or *arithmetic mean*. It is found by adding a set of scores together and dividing by the number of scores.

capacity The amount of liquid that a container can hold. This is measured in millilitres (mL), litres (L), kilolitres (kL) and megalitres (ML). There is a close relationship between capacity and volume.

Celsius scale A scale for measuring temperature. It has 100 equal divisions called degrees. 0 degrees is the temperature at which ice melts and 100 degrees is the temperature at which water boils. The symbol for degree Celsius is °C.

circumference The distance around the edge of a circle.

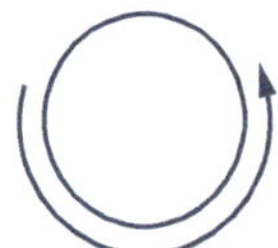

column (or bar) graph A graph that uses horizontal or vertical columns of varying lengths to represent information. The lengths of the columns vary according to the number of items or values that are represented.

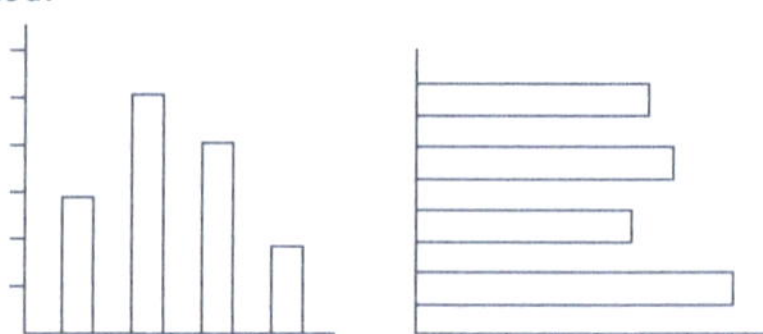

common factor A factor that a set of numbers has in common. For example, the number 3 is a common factor of 6 and 9.

commutative law of multiplication The order in which two or more numbers are multiplied does not affect the answer. For example, $7 \times 4 = 4 \times 7$ and $3 \times 2 \times 5 = 2 \times 3 \times 5$.

comparative Able to be compared. For example, the *comparative cost* of an item is the cost of that item compared to other similar items.

congruent Two shapes are congruent if they have the same size and shape.

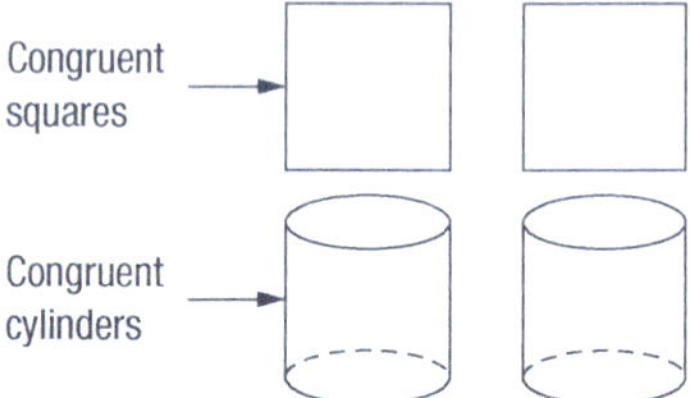

constant A number or symbol that remains the same. To be 'constant' means to 'remain the same'.

continuous data Data that is not separate. Continuous data can take on any value within a range. It is usually any data that is measured because scales of measurement have meaning at all points between the numbers given. For example, ages have meaning whether they are whole numbers or between whole numbers (15 years, 15.6 years, 15.9 years); times have meaning whether they are whole seconds or between whole seconds (10 seconds, 10.27 seconds, 11.5 seconds).
The other type of data is called discrete data.

convert To change something into something equivalent. For example, 1 kilometre converts to 1000 metres; the common fraction $\frac{1}{2}$ converts to the decimal fraction 0.5 and to the percentage 50%.

cube root The cube root of a given number is the number that, when multiplied by itself three times, equals the given number. For example, the cube root of 64 is 4 because $4^3 = 4 \times 4 \times 4$.

cumulative total A total that increases in size or quantity by successive additions. For example, the total of 15 and 27 is 42 and if another 18 is added to this total then the cumulative total is 60.

data Also known as *statistics*. A set of data is information (numbers or words) collected as part of a survey or questionnaire. For example, what colour is most popular among students in our grade? The data set is the number of students who like each colour. The set of data can be shown in a table or on a graph to help us interpret it better. Data can be discrete or continuous.

decimal place The position of a digit in a decimal fraction, that is, tenths, hundredths, thousandths etc. If, for example, a number with three decimal places such as 2.637 is to be rounded to two decimal places then it becomes 2.64 because 7 thousandths is closer to the next hundredth (4) than the previous hundredth (3).

degree A unit for measuring temperature (see Celsius scale) and angles, including latitude and longitude. There are 360 degrees (360°) in a full turn or circle. The symbol for degree is °.

discount An amount of money that is deducted from the total amount that is due. For example, if there is a 20% discount off a total amount of K100, the total amount is reduced by K20 and becomes K80.

discrete data Data that is separate or distinct. Discrete data contains distinct values and is based on counts. Only a certain number of values is possible and the values cannot be subdivided meaningfully. For example, a count of the number of phone calls made each day would be presented as whole numbers (e.g. 5 calls) not as parts of numbers (e.g. 5.25 calls).
The other type of data is called continuous data.

distortion A change in shape or appearance so that the original shape or appearance is misrepresented.

duration The length of time taken for an event to occur. It is also known as *elapsed time* or *time passed.*

elements of a set A group of numbers, shapes or items with a particular thing in common. For example, the numbers 2, 4, 6, 8, 10, 12, 14, 16 and 18 are the set of even numbers less than 20. The things that belong to a set are called *elements*, so the set described has nine elements.

equilateral triangle A triangle that has three equal sides and three equal angles.

error margin It quantifies the degree of accuracy acceptable in a solution. The symbol ± is used to help define the range. For example, 10 ± 5 means solutions within the range 5 to 15 are acceptable.

exponent The number that shows the number of times the base number is multiplied together to give the product. For example, in 2^3, 2 is the base number and 3 is the exponent. 2^3 is read as '2 to the power of 3' or '2 to the third power', and means $2 \times 2 \times 2$ (8). It is also known as *index* or *power.*

factor Any number that you can divide into another number without leaving a remainder, for example, factors of 6 are 1, 2, 3 and 6; factors of 10 are 1, 2, 5 and 10. The smallest factor of a number is always 1 and the largest factor is always the number itself.

factor tree A diagram that shows all the factors of a given number. It can be used to find the prime factors of a number. The factor tree here shows that the prime factors of 240 are 2, 3 and 5.

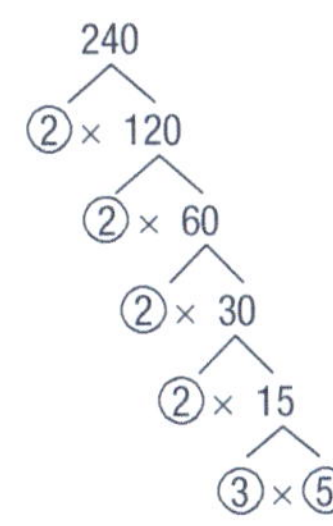

Fahrenheit scale A scale for measuring temperature. On this scale, the freezing point of water is 32 degrees and the boiling point of water is 212 degrees. The symbol for degree Fahrenheit is °F.

formula (*plural*: **formulae**) A rule that shows how to work out an unknown value by using known values. It is commonly expressed in algebraic symbols, for example, the circumference of a circle is twice its radius times pi so the formula is $C = 2\pi r$.

fraction A fraction is part of a group or a whole. It can be written as a common fraction, a decimal fraction, a percentage or a ratio.

1 out of 4 parts = $\frac{1}{4}$

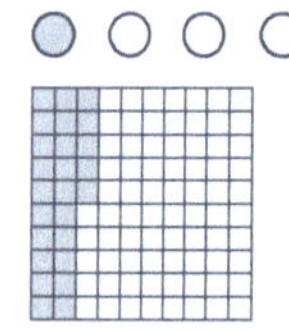

2 out of 8 counters = $\frac{2}{8}$ or $\frac{1}{4}$

25 out of 100 = 0.25 or 25%

Mix 1 part to 4 parts = 1 : 4

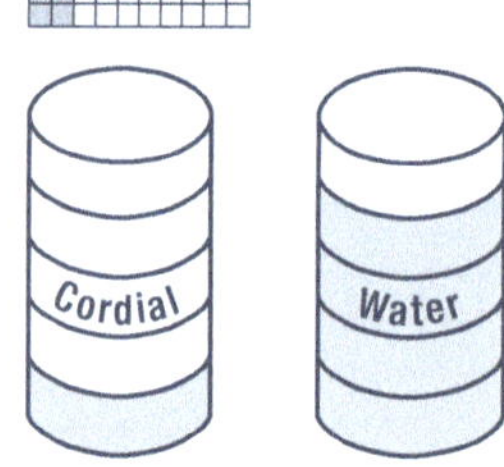

Common fractions are written in the form $\frac{a}{b}$ where a is the numerator and b is the denominator.
A unit fraction is a common fraction that has 1 as its numerator, for example, $\frac{1}{5}$, $\frac{1}{3}$.

fractional indices (indices is the plural of **index**). A whole number index shows how many times to use the number in a multiplication. A fractional index shows how many times to take the root of a number. For example, $4^{\frac{1}{2}}$ means take the square root and is written as $\sqrt{4}$. $8^{\frac{1}{3}}$ means take the cube root and is written as $\sqrt[3]{8}$.

frequency table A graph or table showing how often an event or quantity occurs.

Numbers thrown on dice		
1	~~||||~~ |||	8
2	~~				~~		6
3	~~||||~~ ||||	9
4						4
5	~~				~~		6
6	~~||||~~ ||	7

generalise To extend to a larger group general principles or ideas that have been formed with a select or sample group.

hexagon A polygon that has six straight sides. A regular hexagon has six sides of equal length and six equal angles.

histogram A type of bar graph that has the bars joined up. It shows the frequency of data in each group.

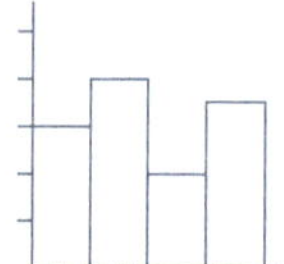

horizontal In the same direction as the horizon. When you are lying down you are horizontal.

humidity (or **relative humidity**) The amount of water vapour in the air. In our daily language the term 'humidity' is generally taken to mean 'relative humidity'. Relative humidity is important in forecasting weather as it indicates the likelihood of rain, dew or fog. Relative humidity is expressed as a percentage. A humidity of 0% means there is no water vapour in the air, while a humidity of 100% means that the air has as much water vapour as it can hold.

imperial measurement A system of measurement that was used before the metric system was introduced. Some countries still use a mixture of imperial and metric measures as the conversion to metric requires education of people, and changes to machinery and production equipment. The United States of America is the only major country that has retained a non-metric system. Some of the units used in the imperial system are inches, feet, yards, miles, ounces, pounds, pints, quarts and gallons.

index (*plural*: **indices**) The number that shows the number of times the base number is multiplied together to give the product. For example, in 4^3, 4 is the base number and 3 is the index. 4^3 is read as '4 to the power of 3' or '4 to the third power', and means $4 \times 4 \times 4$ (64). It is also known as *exponent* or *power*.

index notation A short way of writing large numbers. The index or exponent shows the number of times the base number is multiplied together to give the product. For example, in 5^4, 5 is the base number and 4 is the index. 5^4 is read as '5 to the power of 4' or '5 to the fourth power' and means $5 \times 5 \times 5 \times 5$ (625). Index notation is also known as *exponential notation*.

intersecting set It shows the elements that are common to both sets. For example, this Venn diagram shows two sets. The first set is the set of odd numbers to 20. The second set is the set of numbers to 20 that are divisible by 5. The intersection of the sets shows the numbers that are odd and that can be divided by 5. These numbers (5 and 15) are elements common to both sets.

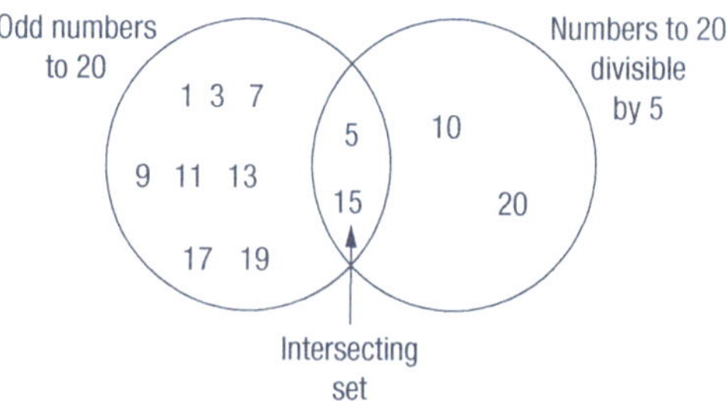

inverse relationship The relationship that reverses the action of the original operation. For example, $4 \times 7 = 28$ and $28 \div 7 = 4$ are the inverse of one another.

key Also known as a *legend*. A list that explains the symbols used on a map, a graph or in a table.

line graph A graph that shows a trend or relationship. It shows how two pieces of information are related and how they vary depending on one another. It can be used to show how something changes over time. Segments of straight lines connect points that represent certain data. A line graph has a vertical axis (y-axis) and a horizontal axis (x-axis).

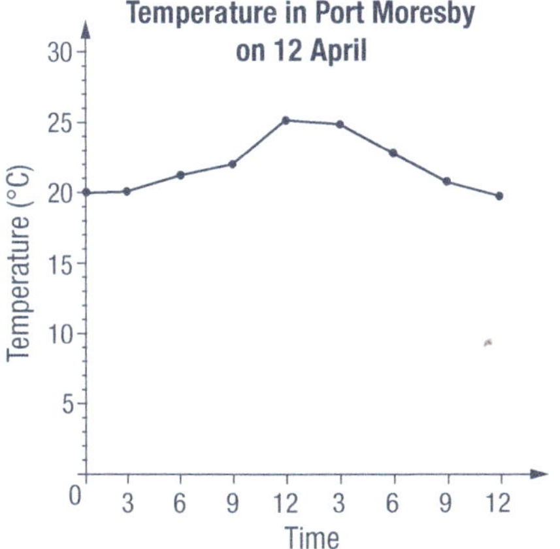

lines of latitude and longitude Lines drawn on a map that can be used together to describe the exact location of a place on the Earth's surface. Lines of latitude are imaginary lines that run in an east–west direction around the Earth. Lines of longitude are imaginary lines that run in a north–south direction from pole to pole. Latitude and longitude are measured in degrees. The lines of latitude show the distance from the equator in degrees. The lines of longitude show the distance from the prime meridian in degrees.

mass The measure of the amount of matter something contains. Mass is measured in grams, kilograms and tonnes. Mass is often confused with weight. The mass of an object remains constant from place to place. Weight is the pull of gravity on an object, and can vary from place to place. For example, an astronaut has the same mass on Earth as on the moon but weighs more on the Earth because the pull of gravity is stronger.

maximum The highest or largest value. For example, the highest temperature for a certain day is referred to as the maximum temperature for that day.

mean It is also called the *arithmetic mean* or *average*. It is found by adding a set of scores together and dividing by the number of scores. For example, a student threw 6, 5, 7, 4, 3 and 5 goals in six different basketball games. The mean is found by adding the number of goals together, $6 + 5 + 7 + 4 + 3 + 5 = 30$, and then dividing by 6 (the number of games), which is $30 \div 6 = 5$. So, the mean of the goals thrown by the student is 5.

median The middle number in a set of numbers when the numbers are arranged in order. If there is no middle number because the number of scores is even, the mean of the two middle numbers is taken. In the set of numbers 3, 5, 8, (12), 13, 17, 26, the median is 12.

metric system A decimal system of measurement. It is a relatively modern system that was developed in France during the 1790s. It is a simple system that uses multiples of 10 to produce larger and smaller units. Common units used in the metric system are centimetres, metres, kilometres, grams, kilograms, millilitres and litres.

minimum The lowest or smallest value. For example, the lowest temperature for a certain day is referred to as the minimum temperature for that day.

minute A unit used both for measuring time and for measuring angles.
As a unit for measuring time: 1 minute = 60 seconds
60 minutes = 1 hour
As a unit for measuring angles: 60 minutes = 1 degree
The symbol for minute as a measure of angle is ′. For example, Port Moresby is located 9°30′ S and 147°10′ E.

mode The most common or frequently occurring number in a set of numbers. For example, in this set of numbers 5 is the most common: 2 5 8 5 3 2 5 5 9 5. There can be more than one mode in a set, and if each number only appears once (or the same number of times) then there is no mode.

null set Also called an *empty set*. It is a set that does not have any elements. For example, the set of shapes that are both triangles and circles is a null set.

parallel Two or more lines or lengths that never meet or cross each other but always stay the same distance apart.

percentage A fraction expressed in hundredths. The symbol for percentage is %. For example, $\frac{27}{100} = 27\%$; $\frac{1}{2} = \frac{50}{100} = 50\%$.

percentage change A number can be increased or decreased by a certain percentage. This is referred to as a percentage change. For example, to increase 35 by 25% we say $35 \times 1.25 = 43.75$; to decrease 35 by 25% we say $35 \times 0.75 = 26.25$. To calculate a percentage change we need to know the original number and the increase or decrease. For example, if the original number is 26 and it has been increased to 38 we put the increase (12) over the original number (26) and multiply by 100: $\frac{12}{26} \times 100 = 46\%$, so the percentage change (increase) is 46%.

percentage error The difference between an estimate and the actual value, expressed as a percentage of the actual value. For example, an estimate is 40 and the actual value is 50. The difference between the two numbers is 10, and 10 as a percentage of 50 is 20%. The percentage error is therefore 20%.

percentile A value on a scale that indicates the percentage of a distribution that is equal to it or below it. For example, a score at the 75th percentile is equal to or better than 75% of the scores; or if 50% of the population were shorter than you, then your height would be said to be at the 50th percentile.

pictogram A graph where the information is shown in pictures or symbols. Each picture or symbol represents a certain number or amount of something. This is shown by a key somewhere on or near the graph. A pictogram is also known as a *picture graph* or *pictograph*.

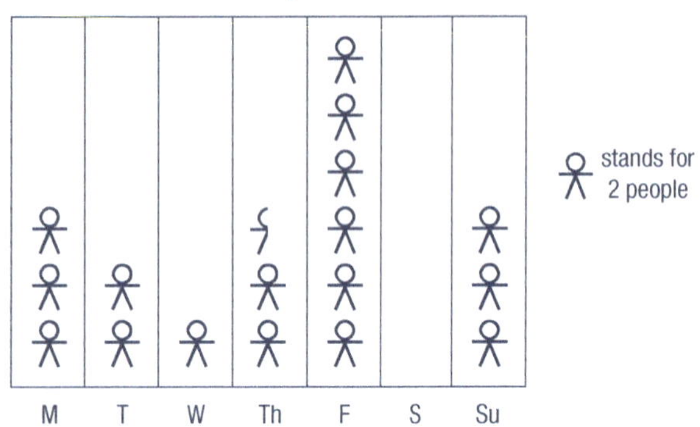

pie chart A graph that is drawn as a circle and shows data as a fraction of the whole, like a pie cut into slices. A pie chart is also known as a *pie graph* or a *circle graph*.

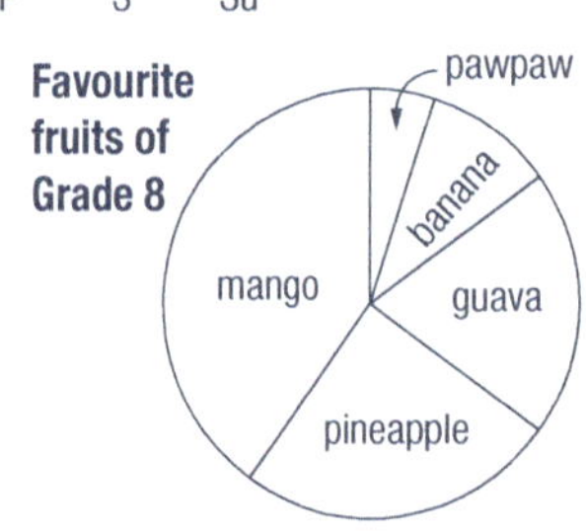

population The number of inhabitants in a given place (country or city etc.). It can also mean a group of items or individuals that share one or more characteristics from which data can be gathered and analysed. For example, all grade 8 students in Papua New Guinea; the group of people who shop at a particular supermarket; all the animals in a village.

pound A unit of mass used in the imperial system of measurement. 1 pound = 0.45359237 kilogram. When converting between kilograms and pounds, we can use the rate 1 kilogram ≈ 2.2 pounds.

predict To state possible results or outcomes about a future event based on current information, collected data or past experience.

prime factor A number that is both a factor and a prime number. For example, the prime factors of 15 are 5 and 3. The prime factors of 12 are 3 and 2.

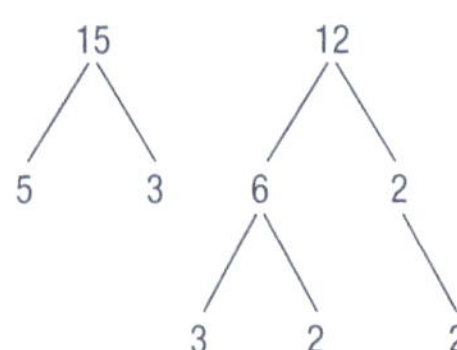

prime number A whole number that has only two factors: itself and 1. That is, it can only be divided by itself and 1 without leaving a remainder. Prime numbers below 20 are 2, 3, 5, 7, 11, 13, 17 and 19. The number 1 is not considered to be a prime number.

product The answer to a multiplication. For example, in $6 \times 7 = 42$, the product is 42; in $2 \times 6 \times 5 = 60$, the product is 60.

profit The amount of money remaining after all costs are deducted from the income of a business.

pronumeral A letter or symbol that stands for an unknown value. For example, in $8 = 5 + y$, the y stands for 3; in $3y - 4 = 17$, the y stands for 7.

protractor An instrument for measuring and marking out angles. Its scale is marked in degrees.

psychrometer A wet-and-dry thermometer that measures the relative humidity.

random sample A group or sample taken from any given population in which every individual in the population being studied has an equal chance of being selected. For example, in a given population of grade 8 at a particular school, a random sample of six students is chosen by drawing their names out of a hat that contains the name of each student in the grade.

range Also called *spread*. It is the difference between the highest and lowest values of a set of data. For example, in the following set of numbers the range is 15 which is the difference between 20, the highest number, and 5, the lowest number: 5, 7, 10, 11, 14, 15, 17, 18, 20.

rate The comparison between two quantities which may be of different things. For example, the rate of water flow from the water tank was 180 litres per hour. This rate compares volume with time.

ratio A comparison of two quantities, measurements or numbers of items. The symbol : is used to express a ratio. For example, when mixing cordial and water in the ratio 1:4, we mix one part cordial to every four parts water.

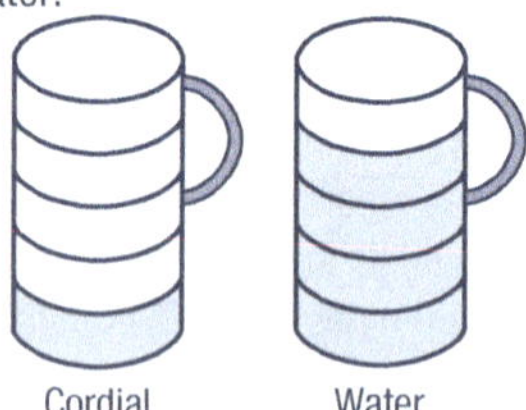

rational number A number that can be expressed as a ratio of whole numbers (positive whole numbers, negative whole numbers and zero). A rational number can be written as a common fraction or a decimal. All whole numbers are rational numbers since they can be written as fractions over 1, for example $5 = \frac{5}{1}$. Other examples of rational numbers are: $\frac{2}{3}$, $-\frac{8}{5}$, 0.5 and 9.

regular shape A shape is regular if all its sides are equal in length and all its angles are equal in size. An equilateral triangle and a square are examples of regular shapes.

relative frequency An approximation of the probability of an event occurring. The relative frequency of a certain value is written as a decimal fraction. For example, a relative frequency of 0.37 means that there is a 37% probability that the given event will occur.

rotate To turn around a centre point or axis.

rotational symmetry When a shape looks exactly the same when turned around a fixed point. The number of times that the shape still looks the same when being turned is the order of symmetry. For example, the shape below has rotational symmetry of order 4.

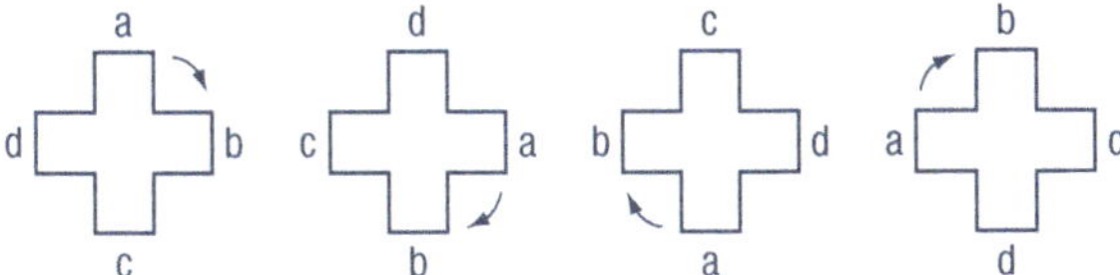

sample A subject chosen from a population for investigation.

scale A series of evenly spaced marks that we use to measure something. The scale on a ruler is marked in centimetres and millimetres, while the scale on a thermometer is marked in degrees to show the temperature. The scale on a map or plan shows how much the map or plan has been reduced or enlarged, for example, 1 cm = 1 km.

scientific notation Also known as *standard index form or powers of 10*. It is a short method for expressing very large numbers. For example, 3×10^8 is scientific notation for the number 300 000 000 ($3 \times 10 \times 10 \times 10 \times 10 \times 10 \times 10 \times 10 \times 10$).
Numbers in scientific notation are made up of three parts.
For 3×10^8:

- 3 is the coefficient which must always be a number larger than or equal to 1 and less than 10
- 10 is the base, which must always be 10
- ^{8} is the exponent (or index or power).

square root A number that when multiplied by itself gives the original number. The symbol for square root is $\sqrt{}$. For example, 7 is multiplied by itself to give 49 ($7 \times 7 = 49$) so the square root of 49 is 7 ($\sqrt{49} = 7$).

statistics Also known as *data*. It is information (numbers or words) that is collected as part of a survey or questionnaire. It can be shown in a table or on a graph to help us interpret it better.

subset A set within a set. All the elements of one set belong to the other set. For example, if Set A includes all the students in your class and Set B includes all the girls in your class, then Set B is a subset of Set A because all the elements of Set B are also in Set A. A subset is defined by a separate boundary.

substitute To replace a pronumeral (a letter or a symbol) with a number. For example, if $y = 5$, the value of $2y + 6$ will be 16.

survey A way of collecting data. It can be done by interview, questionnaire or observation.

table of values A table showing two sets of related numbers. In the table shown here $y = 2x$.

x	1	2	3	4
y	2	4	6	8

tally To count by making a mark to represent each item. This is often shown as a group of five with the fifth mark crossing the other four. 𝍸

temperature How hot or cold something is. It is usually measured in degrees Celsius.

template A pattern that is used as a guide to make a design. A shape can be cut from card and then used as a template by putting it in different positions and each time tracing around its outside edge to make a larger design.

tessellate To fit shapes together in a repeating pattern so that they leave no gaps, for example:

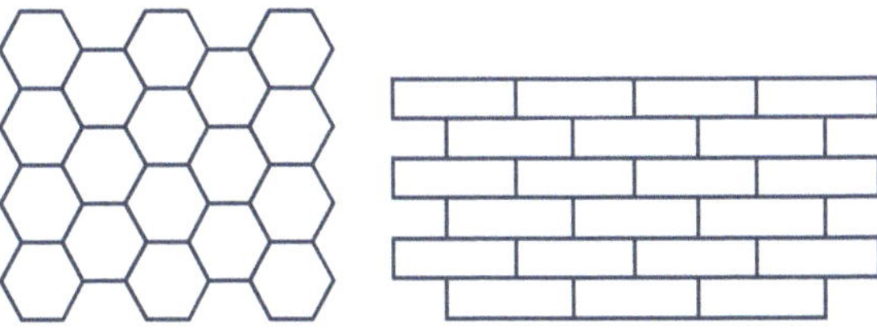

time zone A region of the Earth that has the same time everywhere within it. Most adjacent time zones are exactly one hour apart. There are 24 time zones in the world each bordered by lines of longitude 15° apart.

tonne (t) A large unit of mass. 1000 kilograms make a tonne.

Venn diagram A way of displaying data when items belong in more than one set. It shows the relationship between sets. The places where the circles intersect show the items that belong in both sets.

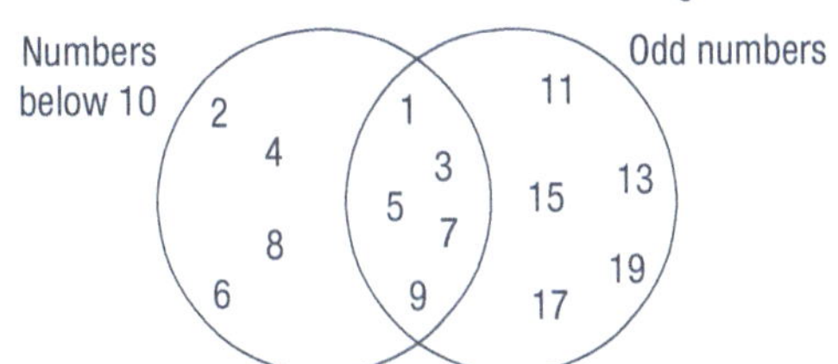

vertical At right angles to the horizon. When you are standing up you are in a vertical position.

volume The amount of space inside a container or the amount of space a solid occupies. The volume of solids is measured in cubic centimetres (cm^3) or cubic metres (m^3). The volume of liquids is measured in litres and millilitres.

weight The gravitational pull on an object. Weight is measured in grams, kilograms and tonnes. Weight is often confused with mass. The mass of an object remains constant from place to place whereas weight can vary from place to place. For example, an astronaut has the same mass on Earth as on the moon but weighs more on the Earth because the pull of gravity is stronger.

x-axis A horizontal number line used to state the position of a point on a graph. It is at right angles to the *y*-axis and the point of intersection, where the axes meet, is called the *origin*.

y-axis A vertical number line used to state the position of a point on a graph. It is at right angles to the *x*-axis. This diagram shows the relationship between the *x*-axis and the *y*-axis.

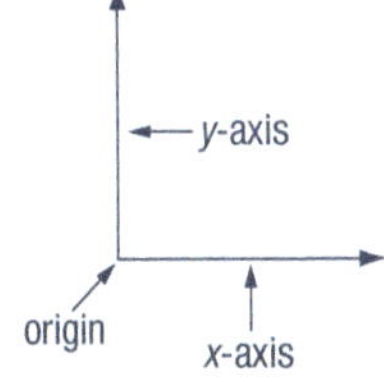

Answers

Topic 3

Learning Unit 1 – Different People, Different Sizes

Lesson 1 – Introduction

1–5 Answers will vary.

Lesson 2 – Weighing in

1 a 96.8 kg b 50 kg

2 a 386.1 kg b 75 kg

3 a 33 kg b 90.9 kg

4 a 70.3 kg b 84.09 kg

5 a 225 kg b 90.9 kg

6 a–b Answers will vary.

7 a–g Answers will vary.

8 Jupiter, Neptune, Earth, Saturn, Venus, Uranus, Mars, Mercury

CHALLENGE: a 1.65 kg b 10 kg

Lesson 3 – Weight for age

1 a about 3.4 kg
  b about 12 months
  c about 5.6 kg
  d From 12 months to 24 months the average girl goes from 9 kg to 11.8 kg (26 pounds ÷ 2.2 = 11.8 kg) which is a gain of 2.8 kg. This is half of what she gains in the first 12 months.
  e A steep line indicates rapid weight gain and as the line becomes less steep the weight gain is less rapid.

CHALLENGE: The greatest weight gain is from birth to 3 months. The line of the graph is steepest from birth to 3 months.

2 a about 3.6 kg
  b about 10.7 kg (14.3 kg – 3.6 kg)
  c 0.2 kg
  d The weight gain from birth to 15 months is about 6.6 kg and from 15 months to 36 months is about 4.1 kg. Weight gain is greater from birth to 15 months than from 15 months to 36 months.
  e about 14.3 kg
  f Answers will vary.

3 250

4 900

5 4 kg (13 kg – 9 kg)

6 A girl on the 5th percentile would gain 9 kg. A girl on the 75th percentile would gain 11.5 kg.

7 a May have health problems – too rapid a weight gain
  b May have health problems – above the 95th percentile at birth and below the 5th percentile at 12 months
  c Normal development
  d May have health problems – gone from the 90th percentile to above the 95th percentile
  e May have health problems – too rapid a weight loss
  f May have health problems – uneven weight gain
  g Normal development
  h May have health problems – has gone from the 50th to the 95th percentile

8–10 Answers will vary.

Lesson 4 – Growing heavier

1 a 7.8 kg b 5.5 kg c 9.6 kg
  d 11.2 kg e 6.2 kg f 9.8 kg

2 a 3 kg b 3.75 kg c 3.375 kg
  d 2.625 kg e 4.688 kg f 4 kg

CHALLENGE:
  a 3.5 kg b 4.38 kg c 3.94 kg
  d 3.06 kg e 5.47 kg f 4.67 kg

3 a 2250 g – 2375 g b 1710 g – 1805 g
  c 2025 g – 2137.5 g d 1755 g – 1852.5 g
  e 2790 g – 2945 g f 2115 g – 2232.5 g

4 a 3:1 b 4:1 c 9:10
  d 95:100 e 3:2 f 13:10

5 a 30 mL b 300 mL c 225 mL
  d 750 mL e 375 mL f 187.5 mL

6 a 4 b 4 c 30
  d 10 e 5 f 60
  g 8 tonic, 16 concentrate, 40 water

7 a 51.75 kg b 47.5 kg c 54.67 kg d 60.6 kg
  e 43.92 kg f 52.5 kg g 50 kg

8 a–b Answers will vary.

CHALLENGE: Answers will vary.

Lesson 5 – Growing taller

1 a 65 cm b 75 cm

2 a about 69.3 cm b about 53.3 cm

3

Birth length	43 cm	44 cm	46 cm	49 cm	52 cm	55 cm
Expected length at 5 months	55.9 cm	57.2 cm	59.8 cm	63.7 cm	67.6 cm	71.5 cm
Expected length at 12 months	64.5 cm	66 cm	69 cm	73.5 cm	78 cm	82.5 cm

4 Graphs will vary depending on the scale chosen. Teacher to check.

5 a 2 b $6\frac{1}{4}$ c $3\frac{1}{8}$ d $7\frac{1}{3}$ e $10\frac{1}{6}$
  f $3\frac{1}{4}$ g $5\frac{1}{4}$ h $6\frac{3}{4}$ i $5\frac{5}{8}$

6 a $4\frac{1}{2}$ kg b $3\frac{3}{4}$ kg c $3\frac{3}{8}$ kg
  d $2\frac{5}{8}$ kg e $4\frac{11}{16}$ kg f 4 kg

7 a 20 b 17 c 17 d 42 e 19
  f 15 g 61 h 20 i 32

8 a 9 b 25 c 11 d 43 e 81
  f 23 g 31 h 17 i 14

9 a–b Answers will vary.

10 a Betty – 158 cm; Mary – 158.5 cm; Jenny – 158.5 cm; Alice – 169 cm

b Otto – 171 cm; Mino – 171.5 cm; Joseph – 171.5 cm; Rodney – 182 cm

Lesson 6 – Body mass index

1 a 26.54 b 31.14 c 24.75 d 25.49 e 28.02

2 a Overweight (pre-obese)
b Answers will vary but must be between 20.0 and 24.9.

3 26.54 – overweight; 31.14 – moderately obese; 24.75 – healthy; 25.49 – overweight; 28.02 overweight

4 a about 15.8 b 23
c overweight d healthy (normal)
e about 17.3 to 24.8 f 26

5 a 15 b about 21 c normal
d healthy (normal) e about 18.8 to 26.3
f about 27.4

6 a 23.67 – overweight b 18.67 – normal
c 13.33 – underweight d 20.47 – normal

7 a–c Answers will vary.

Lesson 7 – Pulse rate and percentage error

1–5 Answers will vary.

CHALLENGE: Answers will vary.

6 a Done
b any answer within the range 15 – 2 and 15 + 2 = (13 to 17)
c any answer within the range 32 – 6 and 32 + 6 = (26 to 38)
d any answer within the range 3.7 – 0.4 and 3.7 + 0.4 = (3.3 to 4.1)
e any answer within the range 11.2 – 0.3 and 11.2 + 0.3 = (10.9 to 11.5)
f any answer within the range 1.1 – 0.05 and 1.1 + 0.05 = (1.05 to 1.15)
g any answer within the range 2.04 – 0.05 and 2.04 + 0.05 = (1.99 to 2.09)

7 a Done
b any answer within the range 120 – 6 and 120 + 6 = (114 to 126)
c any answer within the range 140 – 14 and 140 + 14 = (126 to 154)
d any answer within the range 160 – 8 and 160 + 8 = (152 to 168)
e any answer within the range 45 – 9 and 45 + 9 = (36 to 54)
f any answer within the range 250 – 62.5 and 250 + 62.5 = (187.5 to 312.5)
g any answer within the range 80 – 12 and 80 + 12 = (68 to 92)
h any answer within the range 250 – 15 and 250 + 15 = (235 to 265)

8 a 25% b 4% c 25% d 50%
e 25% f 25% g 14%

Lesson 8 – Estimating measurement

1–5 Answers will vary.

CHALLENGE: Answers will vary.

6–11 Answers will vary.

Lesson 9 – How hot is it?

1 a 37.5°C b 36.9°C c 38.2°C
d 37.1°C e 38.8°C f 37°C

2 38.2°C and 38.8°C – above the upper limit of 37.8°C so indicate a fever

3 a–f Students to draw thermometers. Teacher to check.

4 a Graphs drawn by students will look similar to this:

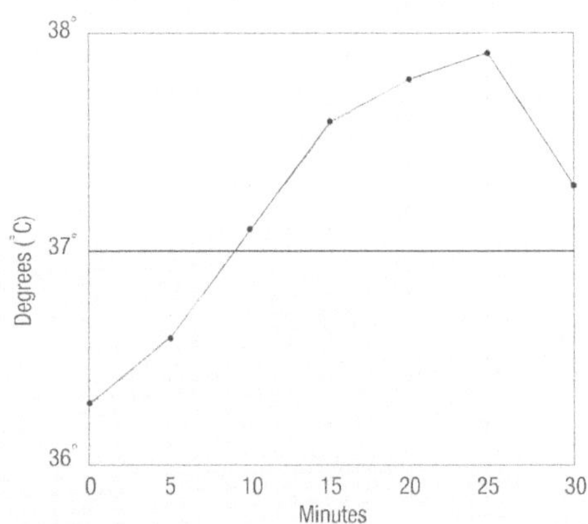

b Students draw two thermometers: one should show 36.3°C and the other should show 37.3°C
c 1°C
d The athlete was outside the healthy temperature range at the 25 minute mark (37.9°C). The upper limit for normal body temperature is 37.8°C.

5 Students to find the relative humidity level.

6–7 Answers will vary.

CHALLENGE: Answers will vary.

Learning Unit 2 – The Past and the Future

Lesson 1 - Introduction

1–4 Answers will vary.

Lesson 2 – How healthy were we then?

1 a Dictionary definitions will vary.
b Answers will vary but should indicate that fewer children per 1000 died in 2006 than in 1990. The rate has improved.

2 Answers will vary but should indicate that for each of the three years women will have a higher average life expectancy than men.

3 Answers will vary but the figures for women should be higher than men.

4 This means that the number of people living in PNG increased from 2000 to 2005 but there was a greater increase in population in the year 2000 than in the year 2005.

5 a More children might get measles.
b Answers will vary.

6 Dictionary definitions will vary.

7 a Infants under 5 years old in PNG and the broader region under which the WHO has classified PNG.
b neonatal causes, other causes and pneumonia
c Answers will vary.
d neonatal causes, other causes and pneumonia
e Answers will vary.
f Answers will vary.
g Graphs will vary depending on the scale used. Teacher to check.

8 a 2100 b 280 c 2520

9 a Answers will vary.
b Papua New Guinea has a higher percentage of HIV/AIDS than the rest of the world.

10 Answers will vary.

11 a–b Answers will vary.

Lesson 3 – Random samples

1 Monday, Friday, Tuesday, Wednesday, Thursday

2 397

3 Week 1 = 37.2; Week 2 = 42.2

4 Answers will vary.

5 No, it would not be reasonable because the data here is for one doctor only.

6 a When this ruler is thrown, 70% of the time it will land numbered side up.
b When this ruler is thrown, it lands numbered side up approximately 50% of the time.

7 a 2500 b 129
c $\frac{15}{129}$ $(\frac{5}{43})$ d about 300

8 a 300 b 50
c 12 d 288

9 a–e Answers will vary.

CHALLENGE: 1666 fish. Explanations will vary but should be similar to this: When the sample of 100 fish was counted, 12 of them were tagged which means about 12% of the population is tagged (12 out of 100). So, if we divide 100 by 12 we get 8.33 and if we multiply this by 200 (the number of fish originally tagged) we get 1666.

Lesson 4 – Making predictions

1 a–d Answers will vary.

2 $\frac{1}{4}$

3 Answers will vary but might include: the location of the clinic; the availability of clinics throughout PNG; the availability of inoculation for babies.

4 a reasonable b not reasonable
c not reasonable d not reasonable

5 b is not reasonable because there are no figures available specifically for grade 8 or grade 10.
c is not reasonable because you cannot generalise to all of Papua New Guinea.
d is not reasonable because 11% of women were over 26 years old (10 from 26 to 30 and 1 over 31).

6–7 Answers will vary.

8 a–e Answers will vary.

CHALLENGE: Answers will vary.

9–10 Answers will vary.

Lesson 5 – Making sense of it

1 a–d Answers will vary.

2 a To show the comparison between the number of cases each year and the cumulative number.
b 1310 c 8770 d 2881

3 a 48 017% b 235 967%

4 a–b Predictions will vary.

5 a tuberculosis, tetanus and measles
b Tuberculosis, because less cases were reported.
c It is not as successful as the other programs.

6 Answers will vary.

7 Graphs will vary depending on the scale chosen.

8 Students to draw a pie chart. Teacher to check.

9–10 Answers will vary.

CHALLENGE: Answers will vary.

Lesson 6 – More on statistics

1 a

Value	Tally	Frequency
15	II	2
16	II	2
17	IIII	5
18	II	2
19	IIII II	7
20	IIII I	6
21	IIII I	6
22	IIII	4
23	I	1
24	III	3
25	I	1

b 19 c 46%

2 a 19.74 b 19 c 20

3 a

Value	Tally	Frequency
48	I	1
49	II	2
53	I	1
56	III	3
58	I	1
60	II	2
61	II	2
64	II	2
71	II	2
72	III	3
85	I	1

b 10 c 50% d 61.9 kg

4 a 8 b 17.4% c 40 – 59

5 15 = 0.12; 16 = 0.18; 17 = 0.32; 18 = 0.22; 19 = 0.12; 20 = 0.04; Totals = 50 and 1.00

6 a 25 – 29 = 0.04; 30 – 34 = 0.09; 35 – 39 = 0.17; 40 – 44 = 0.29; 45 – 49 = 0.28; 50 – 54 = 0.08; 55 – 60 = 0.04

b 29 c 41%

7 a

Value	Tally	Frequency	Relative frequency
5	III	3	0.13
6	IIII	4	0.17
7	IIII II	7	0.3
8	IIII I	6	0.26
9	III	3	0.13

b 39% c 30%

Lesson 7 – Column graphs and histograms

1 a continuous

b

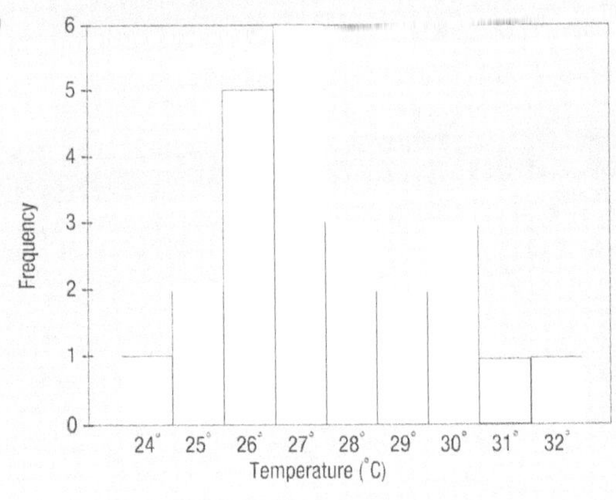

2 a discrete

b

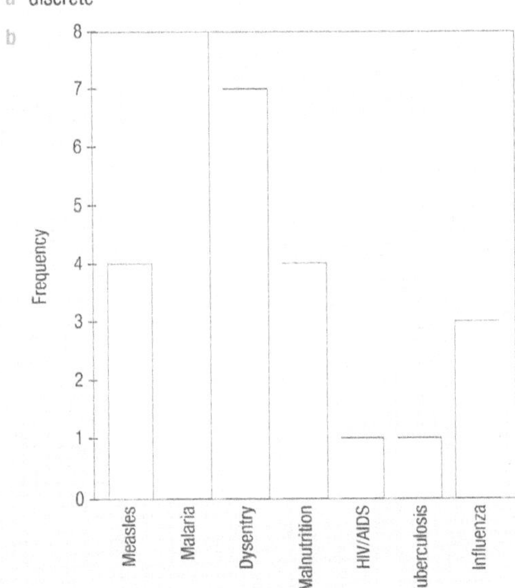

3 a

Weight	Tally	Frequency	Relative frequency
15	I	1	0.04
16	I	1	0.04
17	IIII	4	0.15
18	IIII	5	0.19
19	IIII I	6	0.23
20	II	2	0.08
21	IIII	4	0.15
22	II	2	0.08
23	I	1	0.04

b

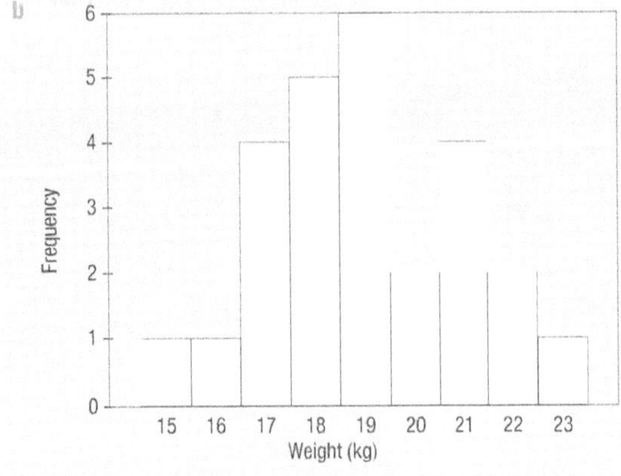

4 a 19 kg b 19 kg c 19 kg

5

Weight (kg)	Frequency
20–29	6
30–39	5
40–49	4
50–59	5
60–69	4
70–79	2

6

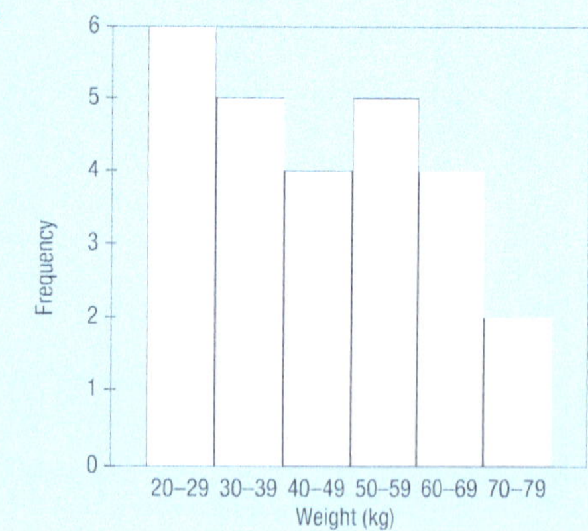

7 a histogram

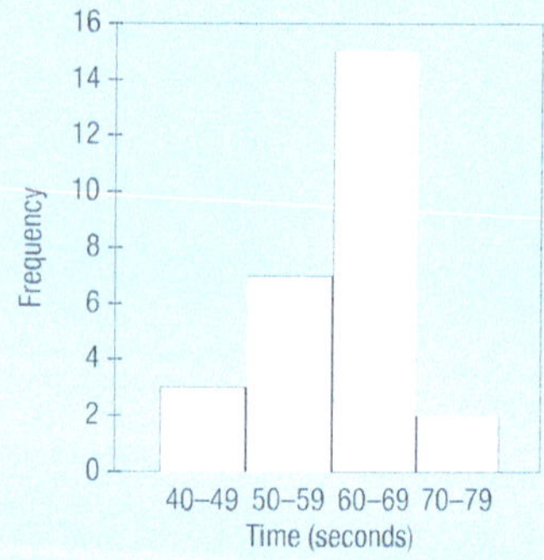

b histogram

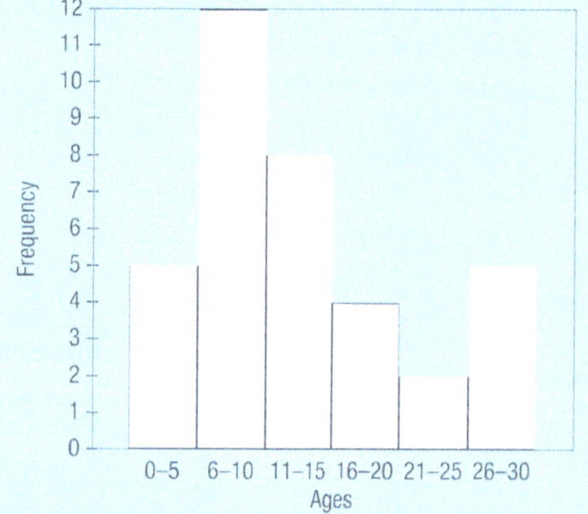

c column graph

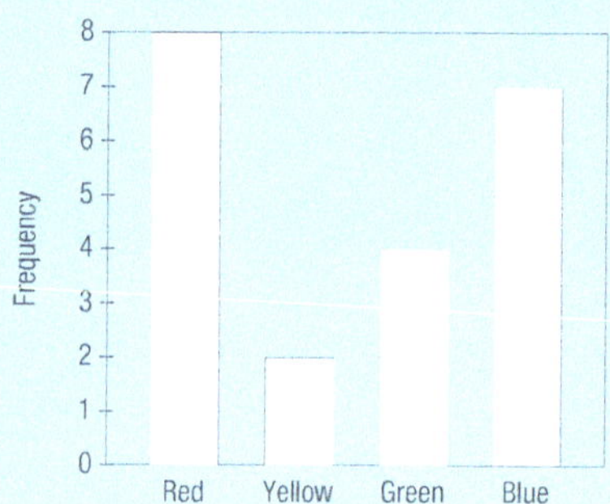

CHALLENGE: Answers will vary.

Lesson 8 – Who belongs where?

1 A = Immunised for measles only
B = Immunised for measles and tetanus
C = Immunised for tetanus only
D = Immunised for measles, tetanus and TB
E = Immunised for measles and TB
F = Immunised for tetanus and TB
G = Immunised for TB only
H = Not immunised

2 Subset = Females under 5 years immunised against tetanus

Intersecting set = Infants under 5 years immunised against measles and TB

Element = One 4-year-old male immunised against TB

Null set = Infants other than males or females under 5 years immunised against TB.

3 a 20 b 10 c 8 d 101 e 5

4 8 patients will receive both injections.

5 a 5 b 13

6 Answers will vary.

CHALLENGE: 20% of the numbers 1–10 are in the set of only odd numbers; 10% of the numbers 1–10 are in the set of only prime numbers; 30% of the numbers 1–10 are in both the set of odd numbers and the set of prime numbers; 40% of the numbers are not in any set.

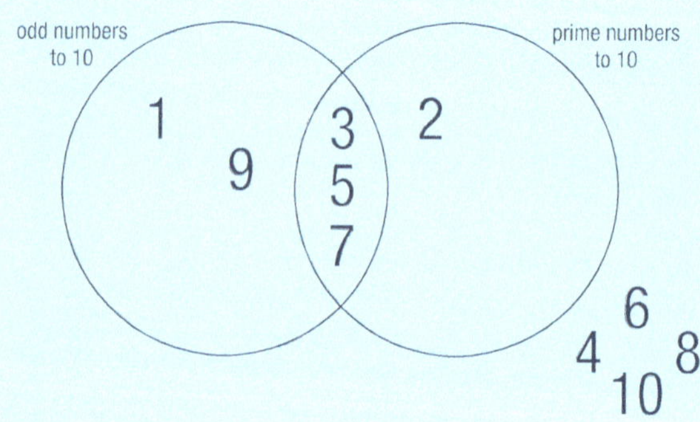

7

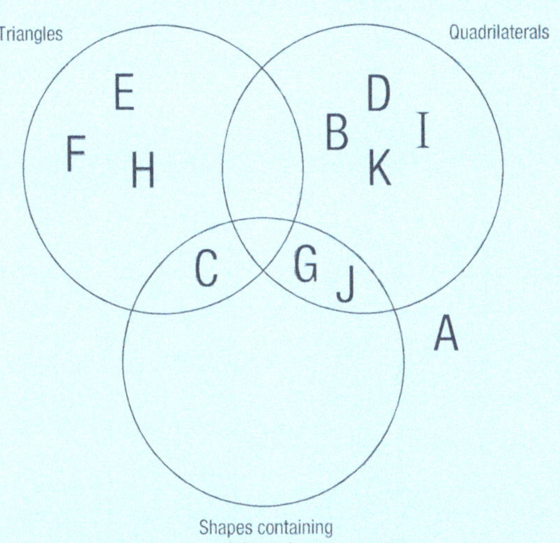

Lesson 9 – Number patterns at work

1 a 64ºF b 50ºF c 68ºF
d 94ºF e 44ºF f 80ºF

CHALLENGE: $C = \frac{F-30}{2}$

2 To convert Celsius to Fahrenheit, multiply the Celsius temperature by 9, then divide by 5 and then add 32.

3 a 59ºF b 71.6ºF c 60.8ºF
d 93.2ºF e 80.6ºF f 51.8ºF

4

Formula	9ºC	13ºC	18ºC	26ºC	29ºC	34ºC
F = 2C + 30	48ºF	56ºF	66ºF	82ºF	88ºF	98ºF
F = 9C/5 + 32	48.2ºF	55.4ºF	64.4ºF	78.8ºF	84.2ºF	93.2ºF

5 a 0.2ºF; 0.6ºF; 1.6ºF; 3.2ºF; 3.8ºF; 4.8ºF

b Answers will vary

6 3 ways are possible:

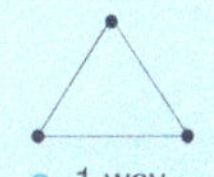

7 a 1 way b 10 ways c 15 ways d 21 ways

8

Nurses	2	3	4	5	6	7	8	9
Paired combinations possible	1	3	6	10	15	21	28	36

9 a 28 days b 10 shifts
c 36 shifts d 21 handshakes

10 a 153 ways b 300 ways
c 406 ways d 435 ways

Learning Unit 3 – Growing and Changing

Lesson 1 – Introduction

1 Graphs will vary depending on the scale chosen. Teacher to check.

2–5 Answers will vary.

6 128

Lesson 2 – A spoonful of medicine

1 Vitamins: 65 240; 26%
Painkillers: 33%
Antibiotics: 17%
Antihistamines: 24%

2 a 25 683 b 24 465 c 1630 d 33 310

3 a 7 mL b 35 mL c 70 mL d 350 mL
e 525 mL f 700 mL

4 a 4 boxes of 100
b 2 boxes of 500 and 2 boxes of 100
c 1 box of 500 and 3 boxes of 100
d 7 boxes of 500 and 1 box of 100

5 a 12 days b 10 mL c 16 days; 10 mL

6 a 20% b 70 cases

7 21 people

8 a 69 people b 2.6%

9 a $\frac{1}{7}$ b 16.7% c 69%

10 a $\frac{1}{7}$ b 35.7% c 2:3 d 45 days

Lesson 3 – Prime time

1 a Students could draw: 1 row of 24; 2 rows of 12; 3 rows of 8; 4 rows of 6; 8 rows of 3; 12 rows of 2; 24 rows of 1

b 1, 2, 3, 4, 6, 8, 12, 24

2 a 1, 3, 7, 21
b 1, 2, 4, 5, 8, 10, 20, 40
c 1, 2, 3, 4, 6, 12
d 1, 2, 3, 4, 6, 9, 12, 18, 36
e 1, 2, 5, 7, 10, 14, 35, 70
f 1, 2, 4, 5, 10, 17, 20, 34, 68, 85, 170, 340
g 1, 2, 3, 4, 6, 9, 12, 18, 27, 36, 54, 81, 108, 162, 243, 324, 486, 972

3 1 = 1; 2 = 1, 2; 3 = 1. 3; 4 = 1, 2, 4; 5 = 1, 5; 6 = 1, 2, 3, 6; 7 = 1, 7; 8 = 1, 2, 4, 8; 9 = 1, 3, 9; 10 = 1, 2, 5, 10; 11 = 1, 11; 12 = 1, 2, 3, 4, 6, 12; 13 = 1, 13; 14 = 1, 2, 7, 14; 15 = 1, 3, 5, 15; 16 = 1, 2, 4, 8, 16; 17 = 1, 17; 18 = 1, 2, 3, 6, 9, 18; 19 = 1, 19; 20 = 1, 2, 4, 5, 10, 20

4 a 1, 2, 3, 5, 6, 10, 15, 30 b 1, 2, 3, 6, 9, 18
c 1, 2, 3 and 6

5 a 1 and 2 b 1, 2, 3, 4, 6 and 12
c 1 and 7 d 1 and 5
e 1, 3, 7 and 21

6 2, 3, 5, 7, 11, 13, 17 and 19

7 a 2, 3 and 7 b 2, 3 and 11 c 3 and 5 d 3, 7 and 11

8 a prime factors = 2, 3 and 5; factors = 1, 2, 3, 4, 5, 6, 10, 12, 15, 20, 25, 30, 50, 60, 75, 100, 150, 300
b prime factors = 2, 3, 5 and 7; factors = 1, 2, 3, 5, 6, 7, 10, 14, 15, 21, 30, 35, 42, 70, 105, 210
c prime factors = 3 and 5; factors = 1, 3, 5, 15, 25, 75, 125, 375
d prime factors = 2, 3, 5 and 7; factors = 1, 2, 3, 4, 5, 6, 7, 10, 12, 14, 15, 20, 21, 28, 30, 35, 42, 60, 70, 84, 105, 140, 210, 420

CHALLENGE: Answers will vary.

Lesson 4 – Back to basics

1 1, 4, 9, 16, 25, 36, 49, 64, 81, 100

2 a 7^3 b 6^5 c 10^4 d 4^2 e m^3

3 a $7 \times 7 \times 7$ b 9×9
c $12 \times 12 \times 12 \times 12$ d 10×10
e $11 \times 11 \times 11 \times 11 \times 11$ f $d \times d \times d$

4 a 3^4 b 8^2 c 5^3 d 2^3
e m^6 f 4^9 g r^2 h b^5

5 a 25 b 16 c 64 d 225
e 10 000 f 1 g 243

6 a $4^2 = 16$ b $6^3 = 216$ c $5^4 = 625$ d $1^{10} = 1$

7 a D b A c B

8 a 2^9 b the same c 3^{10} d 3^4 e 2^6

9 a 1296 b 216 c 400

10 a $(x \times y)^2$ b $(y \times b)^2$ c $(y \times 3)^5$
d $(4 \times x)^{10}$ e $(6 \times m \times p)^3$

11 a 5 cm^2 b 6 mm^2

12 a 2 cm^3 b 10 m^3

CHALLENGE: Answers will vary.

Lesson 5 – How many is that?

1 a 30 b C

CHALLENGE: Answers will vary.

2 a $1 + 3 + 5 + 7 + 9 = 25$ (5^2)
$1 + 3 + 5 + 7 + 9 + 11 = 36$ (6^2)
$1 + 3 + 5 + 7 + 9 + 11 + 13 = 49$ (7^2)
b Descriptions will vary. The sum of the first 9 odd numbers can be worked out by calculating 9^2 (81).

3 a 144 b 625 c 2116 d 10 000

CHALLENGE: Students to carry out an investigation.

4 a $4^2 - 3^2 = 16 - 9$ (or 7)
$5^2 - 4^2 = 25 - 16$ (or 9)
b $6^2 - 5^2 = 36 - 25$ (or 11)
$7^2 - 6^2 = 49 - 36$ (or 13)
$8^2 - 7^2 = 64 - 49$ (or 15)

5 Answers will vary.

6 a 23 b 29 c 945 d 2163

7 a $111^2 = 12\,321$
$1111^2 = 1\,234\,321$
b Answers will vary.

8 a $111\,111^2 = 12\,345\,654\,321$
b $11\,111^2 = 123\,454\,321$
c $1\,111\,111^2 = 1\,234\,567\,654\,321$

9 a 4 and -4 b 8 and -8 c 9 and -9 d 10 and -10
e 11 and -11 f 12 and -12 g 15 and -15

10 a 2 b 3 c 5 d 7 e 10

11 a 6 b 64 c 10 d 121

12 a 2^3 b -5^2 c 3^2 d 2^3 e 3^5

CHALLENGE: Answers will vary.

Lesson 6 – Less than nothing

1

$2^0 = 1$	$5^0 = 1$	$10^0 = 1$
$2^1 = 2$	$5^1 = 5$	$10^1 = 10$
$2^2 = 4$	$5^2 = 25$	$10^2 = 100$
$2^3 = 8$	$5^3 = 125$	$10^3 = 1000$
$2^4 = 16$	$5^4 = 625$	$10^4 = 10\,000$
$2^5 = 32$	$5^5 = 3125$	$10^5 = 100\,000$

2 a 4 b 6 c 10 d 1001 e 26
f 28 g 13 h 66 i 51 j 26

3

Month	1	2	3	4	5	6
Exponent	3^0	3^1	3^2	3^3	3^4	3^5
Calculation	1	1×3	1×3×3	1×3×3×3	1×3×3×3×3	1×3×3×3×3×3
Monthly total	1	3	9	27	81	243

4 a 19 683 b 59 049 c 531 441

5

Month	1	2	3	4	5	6	7	8	9	10	11	12
Exponent	2^0	2^1	2^2	2^3	2^4	2^5	2^6	2^7	2^8	2^9	2^{10}	2^{11}
Total	1	2	4	8	16	32	64	128	256	512	1024	2048

6 a $2^{-3} = 1 \div 2 \div 2 \div 2 = 0.125$
b $2^{-5} = 1 \div 2 \div 2 \div 2 \div 2 \div 2 = 0.031$
c $5^{-3} = 1 \div 5 \div 5 \div 5 = 0.008$
d $10^{-2} = 1 \div 10 \div 10 = 0.01$
e $4^{-1} = 1 \div 4 = 0.25$
f $3^{-2} = 1 \div 3 \div 3 = 0.11$

CHALLENGE: You could use multiplication to check the answers. If the result is 1 then the answer is correct. For example: $0.125 \times 2 \times 2 \times 2 = 1$

7 a 16 b 0.0001 c 1 d 1 e 625
f 0.0625 g 100 h 0.0001 i 1 j 49
k 0.25 l 0.037

8 a $\frac{1}{5^3}$ b $\frac{1}{8^1}$ c $\frac{1}{4^2}$ d $\frac{1}{10^3}$ e $\frac{1}{6^2}$
f $\frac{1}{4^3}$ g $\frac{1}{6^1}$ h $\frac{1}{5^7}$ i $\frac{1}{2^4}$ j $\frac{1}{3^7}$
k $\frac{1}{2^3}$ l $\frac{1}{7^7}$ m $\frac{1}{m^p}$ n $\frac{1}{5^p}$ o $\frac{1}{x^y}$

9 a 2^{-3} b 6^{-2} c 5^{-10} d 4^{-3} e 10^{-4}
f 8^{-2} g 5^{-3} h 9^{-4} i 7^{-2} j 4^{-8}
k r^{-2} l s^{-3} m p^{-4} n 7^{-m} o x^{-y}

CHALLENGE:
a 0.125 b 0.028 c 0.0000001024
d 0.016 e 0.0001 f 0.016
g 0.008 h 0.00015 i 0.02
j 0.000015

Lesson 7 – The spread of disease

1 a 250 b 500 c 1500 d 3000

2 a 2.5 minutes b 3 minutes c about 3 m 40 s

3 10 000

4 $\frac{1}{1000}$

5

Exponent form	Number of cells	Sum of zeros in basic number
10^1	10	1
10^2	100	2
10^3	1000	3
10^4	10 000	4
10^5	100 000	5
10^6	1 000 000	6
10^7	10 000 000	7
10^8	100 000 000	8
10^9	1 000 000 000	9
10^{10}	10 000 000 000	10

6 a $(3^2)^3 = (3 \times 3) \times (3 \times 3) \times (3 \times 3) = 3^6$
b $(2^3)^2 = (2 \times 2 \times 2) \times (2 \times 2 \times 2) = 2^6$
c $(4^2)^2 = (4 \times 4) \times (4 \times 4) = 4^4$
d $(5^3)^3 = (5 \times 5 \times 5) \times (5 \times 5 \times 5) \times (5 \times 5 \times 5) = 5^9$
e $(m^2)^3 = (m \times m) \times (m \times m) \times (m \times m) = m^6$
f $(p^2)^2 = (p \times p) \times (p \times p) = p^4$

7 Answers will vary but should mention that the two index numbers can be multiplied together to give the answer. For example: in $(3^2)^3$, the index numbers 2 and 3 can be multiplied to give 3^6.

8 a D b C c A d A

9 a 3 b 2 c 10 d $\sqrt{m}$ e $\sqrt{x}$

CHALLENGE: 3

Lesson 8 – Health check

1–3 Answers will vary.

4 a Answers will vary b 11.9 L/min
c 31.2 L/min d 138.6 L/min

5 Answers will vary but should indicate that the numbers in the calculation will decrease.

6 a 4 min b 115 L/min c 4 min d 35 L/min

7 Answers will vary.

8 a 100 b 95 c 15.79% d 75 e $\frac{4}{25}$

9 a 0.86% b 7.9%
c From 1995 to 1998 it increased by 6.1%.
d Answers will vary.

CHALLENGE: Students will find that 1 part of K100 is K2.86; 1 part of K5000 is K142.86; 1 part of K800 000 is K22 857.14; 1 part of K3 000 000 is K85 714.29.

In the ratio 9:16:10 the following amounts would be spen

Amount	Health	Education	Defence
K100	K25.74	K45.76	K28.60
K5000	K1285.74	K2285.76	K1428.60
K800 000	K205 714.26	K365 714.24	K228 571.40
K3 000 000	K771 428.61	K1 371 428.64	K857 142.90

Learning Unit 4 – Additional Learning, Revision and Assessment

Lesson 1 – Additional learning

1

Index form	Number	Zeros in the number
10^1	10	1
10^2	100	2
10^3	1000	3
10^4	10 000	4
10^5	100 000	5
10^6	1 000 000	6
10^7	10 000 000	7
108	100 000 000	8
10^9	1 000 000 000	9
10^{10}	10 000 000 000	10

2 a 10^2 b 10^4 c 10^6 d 10^8

3 a 10^2 b 10^3 c 10^4 d 1000; 10^3
e 100 000; 10^5 f 1 000 000; 10^6

4 a 300 b 5000 c 10 000; 60 000
d 100 000; 800 000 e 10 000 000; 20 000 000
f 1 000 000; 1 000 000

5 9 000 000 000 000 000

6 2 000 000 000 000 000 000 000 000 000 000

7 4×10^4

8 55×10^6

9 9 460 000 000 000 000

10 1 989 100 000 000 000 000 000 000 000 000

11 a 15 000 000 b 3500 c 185 000
d 756 000 e 36 000 000 000 f 603 000 000
g 2 850 000 h 13 020 000 000

12 a 10^7 b 10^4 c 10^4
d 10^5 e 10^6 f 10^6

CHALLENGE: Three possible ways are: $4.56 \div 10^3$; $45.6 \div 10^4$; $456 \div 10^5$

Lesson 2 – Revision

1 a 28.5 kg b 62.07 kg

2 a 2.5 kg: 2.25 kg, 3.25 kg, 3.75 kg, 5 kg
b 3200 g: 2880 g, 4160 g, 4800 g, 6400 g
c 2.75 kg: 2.475 kg, 3.575 kg, 4.125 kg, 5.5 kg

3 a 2430 g to 2565 g b 2160 g to 2280 g
c 2142 g to 2261 g d 1.53 kg to 1.615 kg
e 2.34 kg to 2.47 kg f 1.845 kg to 1.948 kg

4 a 3 b 6 c 30 d 20
e 7 tonic, 21 milk powder, 35 water
f 9 g 45

5 a 45.25 kg b 38.67 kg c 44.4 kg d 58.25 kg

6 a 24.31 b 24.22 c 15.11

7 overweight; normal; normal

8 a any answer within the range 135 to 165
b any answer within the range 152 to 168
c any answer within the range 665 to 735
d any answer within the range 52 to 78
e any answer within the range 90 to 150

9 a 38° b 36.5° c 37.8° d 38.7°

10 a April b 38 c 6.33
d Answers will vary.
e No, it would not be reasonable. Reasons will vary.

11 a 300 b 50 c $\frac{48}{50}$ $(\frac{24}{25})$ d 4%

12 a Graphs will vary depending on the scale chosen. Teacher to check.
b Ronnie – 25.8; Amos – 17.4
c Answers will vary.
d Answers will vary.

13 a

Age	Frequency	Relative frequency
4	6	0.15
5	9	0.225
6	6	0.15
7	5	0.125
8	2	0.05
9	3	0.075
10	4	0.1
11	4	0.1
12	1	0.025

b 5 c 6.975 d 5 e 6 f 65%

14 Answers will vary but should include information from page 34 of this book.

15 Eight patients will receive both injections.

16 a 68°F b 89.6°F c 64.4°F d 75.2°F e 57.2°F

17 a 32 – 1, 2, 4, 8, 16, 32
b 50 – 1, 2, 5, 10, 25, 50
c 24 – 1, 2, 3, 4, 6, 8, 12, 24
d 49 – 1, 7, 49
e 64 – 1, 2, 4, 8, 16, 32, 64
f 360 – 1, 2, 3, 4, 5, 6, 8, 9, 10, 12, 15, 18, 20, 24, 30, 36, 40, 45, 60, 72, 90, 120, 180, 360
g 1024 – 1, 2, 4, 8, 16, 32, 64, 128, 256, 512, 1024

18 a 1, 2, 4 and 8 b 1, 2 and 4 c 1 and 11
d 1, 2, 7 and 14 e 1, 2, 3, 4, 6 and 12

19 a 5 b 2, 3 and 17
c 2, 5 and 7 d 2, 5 and 11

20 a 3^2 b 4^5 c 2^{10} d 3^5 e 2^1

21 C

22 a 3^3 b -5^3 c $\sqrt[3]{125}$ d 4^3 e 2^5

23 a 5 b 11 c 33 d 9.0001

24 a $\frac{1}{3^2}$ b $\frac{1}{5^3}$ c $\frac{1}{y^2}$ d $\frac{1}{5^a}$ e $\frac{1}{x^y}$

25 a 6^{-2} b 4^{-3} c h^{-2} d 5^{-p} e x^{-y}

26 1 000 000 cells

27 $\frac{1}{10^2}$

28 a $3.5 \times 10\,000 = 35\,000$
b $6.13 \times 100\,000 = 613\,000$
c $1.75 \times 1000 = 1750$
d $3.06 \times 10\,000\,000\,000 = 30\,600\,000\,000$

29 a 1.5×10^5
b–c Answers will vary.

Lesson 3 – Assessment Task 1

Answers will vary (to be assessed by teacher).

Lesson 4 – Assessment Task 2

1 a 2.25 kg – 2.375 kg b 1710 g – 1805 g
c 2025 g – 2137.5 g

2 a 3.7 kg b 3.256 kg c 3.7 kg

3 a 3 cups b 1.5 cups c $2\frac{2}{3}$ cups

4 a 30–34 b 2.8–4.2 c 1.05–1.15
d 12–18 e 51–69 f 237.5–262.5

5 a 67% b 11% c 50%

6 36.77°C

7 a 96.8°F b 78.8°F c 50°F d 26.6°F

8 a M = 36 b M = 105 c M = 253 d M = 406

9 a Vitamins = 53%; Painkillers = 30%; Antibiotics = 6%; Antihistamines = 11%
b 48
c 20 packets

10

Score	Frequency	Relative frequency
70	2	0.1
72	3	0.15
75	3	0.15
76	1	0.05
77	1	0.05
83	2	0.1
84	2	0.1
86	2	0.1
89	2	0.1
90	2	0.1

11 a 10 b 25% c 79.9 d 80

12 a The number of children under 10 years old diagnosed with either hookworm, measles, diarrhoea or malaria at the Feelgood Clinic in June 2008.
b diarrhoea, hookworm, measles, malaria
c Answers will vary.
d Answers will vary but students should expect there to be 4 or 5 cases of hookworm diagnosed in each of the two weeks beyond the graph.
e No, it is not appropriate as this data is only for one month and from one clinic.

13 a

Classes	Frequency
10–19	6
20–29	7
30–39	7
40–49	3
50–59	3

b Histograms should be similar to the one shown here:

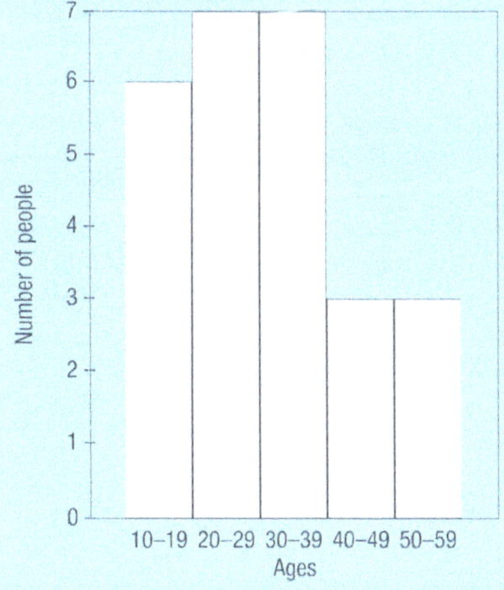

c Answers will vary.

14 a 28 b 16 c 12 500 d 11

15 a $\frac{1}{3^2}$ b $\frac{1}{5^2}$ c $\frac{1}{4^2}$

16 a 8^{-2} b 6^{-3} c 5^{-6}

17 a 450 000 b 5 230 000 c 30 200 000 000

18 a–b Answers will vary.

Topic 4

Learning Unit 1 – Where do Tourists go?

Lesson 1 - Introduction

1–3 Answers will vary.

Lesson 2 – Attractions of Papua New Guinea

1 a Every 1 cm on the map equals 10 000 000 cm in real life.
b $\frac{1}{10\,000\,000}$

2 a Smaller b $\frac{1}{20\,000\,000}$

3 a Larger b $\frac{1}{25\,000\,000}$

4 100 kilometres

5 a 2 cm b 5 cm c 12 cm d 8 ½ cm e 4 ½ cm
f 3.25 cm g 2.75 cm h 9.2 cm i 10.75 cm j 7.8 cm

6 a 250 km b 420 km c 870 km d 580 km e 675 km
f 1025 km g 790 km h 310 km i 1275 km j 125 km

7 The following distances are approximate.
a 350 km b 250 km c 300 km d 175 km
e 500 km f 375 km g 1025 km h 700 km

8 Approximately 1575 km

9 a–c Answers will vary.

10 a 1 cm = 150 km b 1 cm = 2.75 km c 1 cm = 500 m
d 1 cm = 1 km e 1 cm = 65 km f 1 cm = 75 m

11 a 1:5 000 000 b 1:300 c 1:1 250 000
d 1:5000 e 1:2 500 000 f 1:15 000

CHALLENGE: Students should have drawn a rectangle that measures 5 cm by 7 cm.

12 a Students to write 2 of the following: Vanimo, Aitape, Wewak, Kavieng and Port Moresby
b Diving, fishing or surfing
c Kokoda, Goroka, Mt. Wilhelm, Mount Hagen and Mt. Talawe
d Kavieng, Hoskins, Madang, Lae, Lorengau, Milne Bay and Arawa
e Diving, surfing and trekking

13 a 6 b 8

14 The following distances are approximate.
a 5.5 km b 8.25 km c 2.75 km d 6.2 km e 9.6 km

15 Approximately 15 km

16 Approximately 27 km

Lesson 3 – Flying in Papua New Guinea

1 a Lorengau is about 800 km north of Port Moresby.
b Rabaul is about 775 km north-east of Port Moresby.
c Alotau is about 350 km south-east of Port Moresby.
d Mendi is about 525 km north-west of Port Moresby.

2 a Kerema b Daru c Kerema
d Alotau e Hoskins f Tufi

3 a 360 km b 575 km c 520 km
d 225 km e 687½ km

4 a Rabaul b Mendi or Mount Hagen
c Madang d Madang or Kundiawa

5 a 59 minutes b 1 hour and 33 minutes
c 3 hours and 22 minutes d 2 hours and 32 minutes
e 1 hour and 41 minutes f 2 hours and 5 minutes

6 a 1 hour and 22 minutes b 1 hour and 13 minutes
c At approximately 4 pm d 11:20 am

CHALLENGE: The formula is: Speed = distance ÷ time

A plane would be travelling at a speed of 729 kph if it takes 1.75 hours to travel the distance of 1275 km between Daru and Buka Island.

Lesson 4 – The cost of airfares

1 a K217 b K523 c K835

2 Rabaul to Alotau is not a direct flight. We know this because it is circled and circled fares indicate that the flight is not direct.

3 a K308 b K345.40 c K815.10 d K765.60
e K827.20 f K424.60 g K773.30 h K267.30
i K788.70 j K743.60

4 a K33.41 b K26.11 c K66.61 d K44.81
e K37.51 f K41.81 g K28.91 h K47.71

5 a K85.60 b K60.40 c K176.40 d K116.50
e K48 f K40.20 g K53.10 h K130.75
i K59.10

6 Estimates will vary but should be similar to the following:
a K32 b K72 c K86 d K88 e K51
f K27 g K130 h K147 i K70

7 K1698.40

8 a K969.10 b No c K152.90

9 a K231
b Adult = K338.80; Child = K254.10

10 K2009.70

Lesson 5 – Where will we stay?

1 a Yes b Yes c No d Yes
e No f Yes

2 a K24 b K12 c K9.50 d K14.60 e K22.40

3 a K180; K18 b K215; K21.50 c K350; K35
d K190; K19 e K155; K15.50 f K145; K14.50

4 K123.50

5 a VAT = K9.50; Total cost = K104.50
b Cost before VAT = K240; Total cost = K264
c Cost before VAT = K375; Total cost = K412.50
d VAT = K27.50; Total cost = K302.50
e VAT = K16; Total cost = K176
f Cost before VAT = K415; Total cost = K456.50

6 a K341 b K799.70 c K668.80 d K440

7 a K195.25 b K334.40 c K399.85

8 a K146.67 b K239.07 c K282.70

9

	Standard	Premium	Executive
Single	K248	K451.20	K545.60
Twin (Double)	K284	K486.40	K581.60
Triple	K320	K521.60	K616.80

10 a K462 b K1254 c K3520 d K2090
e K5280 f K7392 g K2816 h K9856

11 a 7 nights b K44

12 a K418 b K104.50

13 a K821.33 b K117.33

Lesson 6 – What will we do?

1

	10% VAT	Total cost
1	K104	K1144
2	K104	K1144
3	K104	K1144
4	K96.50	K1061.50
5	K96.50	K1061.50
6	K124.80	K1372.80
7	K124.80	K1372.80

2 a K1259.80 b K13855.60

3 Answers will vary.

4 a K69 278 b K6298

5 a K10 194.80 b K1132.76 c K102.98

6 Company B

7 K94 500

8 a K128 160 b K13 500

9 a K17 850 b Answers will vary.

10 Company X = K2510 more; Company Y = K1910 more

11 Answers will vary.

12 K11 475

13 a K93 060 b K68 850

14 K64 400

CHALLENGE: Answers will vary.

15 a–b Answers will vary depending on the cabins and companies that students choose.
c K46 206.40

Lesson 7 – Patterns in handcrafts

1 Shapes A, B, D, E and F have rotational symmetry.

2 a order 2 b order 4 c no rotational symmetry
d order 3 e order 6 f order 5

3 Shape C

4 a–d Answers will vary.

5 Answers will vary.

6 Students to make a rotational tessellation.

7 a–c Students construct three shapes.

8 Shapes will vary.

9 Designs will vary.

10 Answers will vary.

11 a–b Answers will vary.

12 Answers will vary but reasons given should mention that the basic shape is either a square, equilateral triangle or regular hexagon that has been changed according to the rules for making rotational tessellations.

Lesson 8 – Renting a car

1 a 36 L b 10.4 L c 20.235 L d 49.47 L
e 28.48 L f 15.265 L g 15.708 L h 37.344 L

2 a K99.72 b K28.81 c K56.05 d K137.03
e K78.89 f K42.28 g K43.51 h K103.44

3 Answers will vary.

4 a 20 L b 14 L c 45 L
d 30 L e 35 L f 35 L

5 a 15 L b 45 L c 44 L
d 16 L e 24 L f 14 L

6 a 8.9 L b 8 L c 8.8 L
d 6.8 L e 5.9 L f 6.8 L

CHALLENGE: 326 km

7 K732.90

8 a K1435.70 b K287.14

9 a K7738 b Answers will vary.

Lesson 9 – Planning a journey

1 a 4 places b 3 nights
c 3 flights within Papua New Guinea d 7 nights

2 a–b Answers will vary.

3 a–c Answers will vary.

4 a 975 m
b Arrows are marked to show tourists which way to walk.
c south-west d 3.275 km e About 4 hours

5 Answers will vary.

6 a–b Answers will vary.

Lesson 10 – Holiday or work?

1 a 2005 b 2002 c 9368
d 4284 e 2305 f 27 919

2 Answers will vary.

3 a 28.1% b 25.7% c 30.7%
d 26.9% e 32% f 33.5%

4 a 1997 b 1999 c 2004

5 a 1997 b 2001 c 2005

6 1996 and 2000

7 a 7.9% b 10.5% c 27.3%

8 a 15.1% b 33.3% c 15.6%

9 a 7.2% b 13.8% c 20.4%

10 a 17.7% b 9.8% c 5.2%

CHALLENGE: Students to draw a graph. Graphs may vary.

Learning Unit 2 – Travelling Overseas

Lesson 1 - Introduction

1–7 Answers will vary.

Lesson 2 – Leaving Papua New Guinea

1 a Tropic of Cancer, Equator, Tropic of Capricorn
b Tropic of Cancer c Tropic of Capricorn

2 Three of the following cities: Brisbane, Sydney, Melbourne, Auckland and Johannesburg.

3 Three of the following cities: Hong Kong, Manila, Bangkok, Calcutta, Karachi and Lagos.

4 a Between the Equator and the Tropic of Capricorn
b North of the Tropic of Cancer

5 Because the following distances are approximate answers may vary but should be within reasonable range of the given distances.
a ≈ 1710 km b ≈ 4560 km c ≈ 9120 km
d ≈ 13 680 km e ≈ 10 260 km

6 a ≈ 6840 km b ≈ 5130 km c ≈ 3135 km
d ≈ 3990 km e ≈ 7980 km f ≈ 14 535 km
g ≈ 10 260 km h ≈ 6840 km

7 a True b False c True
d True e False f True

8 18 240 km

9 13 965 km

10 Answers will vary.

Lesson 3 – Where in the world is …?

1 a–b Answers will vary.

2 a–b Answers will vary.

3 a–e Answers will vary.

4 a Iceland
b–c Answers will vary.

5 a–e Answers will vary.

6 a Aoba b Port Vila c Ranon d Merig Island

7 a 16° 5′S, 167° 25′E b 16° 40′S, 168° 10′E
c 16° 25′S, 167° 50′E d 15° 20′ S, 166° 40′E
e 15° 30′S, 167° 10′E f 13° 25′S, 166° 40′E

8 a Suva b Nadi c Wairiki d Nabouwalu

9 a 17° 40′S, 178° 50′E b 17° 25′S, 178° 15′E
c 19° 5′S, 178° 15′E d 18° 10′S, 177° 30′E
e 17° 35′S, 177° 30′E f 16° 25′S, 179° 25′E

CHALLENGE: Answers will vary.

Lesson 4 – Temperature variations

1 a Answers will vary but the places chosen should be near the Equator.
b Answers will vary but given reasons should include that the places are close to the Equator where the sun is high overhead all year.
c Answers will vary but the places chosen should be near the poles.
d Answers will vary but given reasons should include that the places are close to the poles where the sun is not high overhead all year.

2 a Nairobi, Jakarta and Kuala Lumpur because there is little variation in average temperatures.
b Vancouver, Moscow and London because there is a large variation in average temperatures.

3 a July and August
b January, February and March
c clothes for cold and warm weather
d clothes for warm to hot weather

4 a August
b December and January
c All months except maybe July and August

5

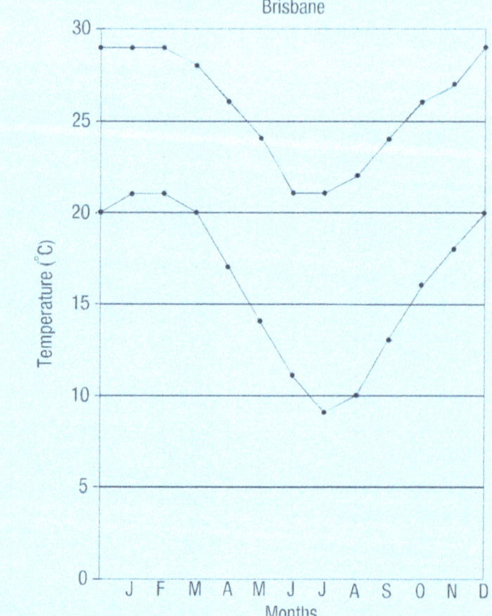

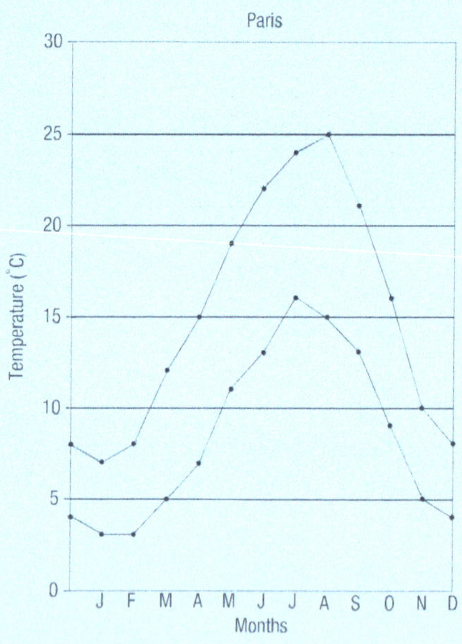

6 a The first graph (Brisbane) curves downwards and the second graph (Paris) curves upwards.
b Brisbane
c December, January and February in Paris are the coldest months.
d–e Answers will vary.

7 a 49°N
b 27°S or 28°S

8 a Tokyo and Quebec because they experience winter in the same months as places in the southern hemisphere experience summer.
b Melbourne and Buenos Aires because they experience winter in the same months as places in the northern hemisphere experience summer.
c Quebec

9 a–d Answers will vary.

10 Answers will vary but should include the fact that records of temperatures need to be kept for many years.

Lesson 5 – Time zones

1 a 10 a.m b 8 a.m. c 12 noon d 4 p.m.
e 7 p.m. f 9 a.m. g 7 p.m. h 6 p.m.

2 a 9 a.m. b 11 p.m. c 10 a.m. d 4 p.m.
e 2 p.m. f 10 a.m. g 7 a.m. h 2 a.m.

3 a 8 a.m. b 11 p.m. c 8 p.m. d 3 p.m.
e 8 a.m. f 1 p.m. g 5 a.m. h 9 p.m.

4 a 8 a.m. b 2 p.m. c 5 a.m. d 6 a.m.
e 9 p.m. f 3 p.m. g 1 p.m. h 9 p.m.

5 a 7 p.m. b 1 a.m. c 11 p.m.

6 a 8 a.m. b 8:25 a.m.

7 a 11 a.m. b 12:30 a.m. c 10:30 a.m.

8 a 6 a.m. b 3 p.m. c–d Answers will vary.

9 a Thursday b Wednesday c Thursday
d Wednesday e Thursday f Thursday
g Thursday h Wednesday

10 a 9 a.m on May 15th b 4 p.m. on May 14th
c 12 noon on May 15th d 7 p.m. on May 14th

CHALLENGE: Ruth celebrated her birthday twice because she had her birthday on July 12th and then on July 13th she crossed the International Date Line in an easterly direction where the date was again July 12th.

Lesson 6 – Plane specifications

1 a 1245 b 1200 c 1040
d 990 e 2850 f 2325

2 225

3 a 21.82 m b 25.8 m c 25.88 m d 32.83 m

4 a 25.91 m b 46.89 m c 26.45 m d 27.39 m

5 362.85 m

6 a 119 360 L b 1 301 040 L c 380 709 L
d 775 200 L e 204 510 L f 1 706 720 L

7 a 6 hours b 5 hr 12 min c 7 hr 45 min
d 2 hr 15 min e 1 hr 24 min f 7 hr 20 min

8 a 1750 km b 4840 km c 7875 km
d 3115 km e 3612.5 km f 2310 km

9 a 461 L b 622 L c 1612 L
d 1048 L e 602 L f 894 L
g 2042 L

10 a Boeing 737 b Airbus 380

11 Boeing 727 = 4 hr 31 min; Boeing 737 = 7 hr 32 min; Boeing 747 = 14 hr 56 min; Boeing 757 = 8 hr 29 min; Boeing 787 = 15 hr 47 min; Airbus 320 = 5 hr 43 min; Airbus 340 = 16 hr 49 min; Airbus 380 = 16 hr 53 min

12 a Boeing 747; Boeing 787; Airbus 340; Airbus 380
b Boeing 727; Airbus 320
c Boeing 737; Boeing 757

13 a–b Answers will vary.

14 a Airbus 380 b Boeing 737 c Boeing 747
d Boeing 747 e Airbus 320 f Boeing 737

CHALLENGE: Questions will vary.

Lesson 7 – Pack your bag

1 Answers will vary but the dimensions must total 115 cm.

2 a No b Yes c Yes

3 a Yes b No c Yes

4 Teacher to check.

5 First class = 24 kg; Business class = 34 kg; Economy class = 26 kg

6 a 3674.55 kg b 1325.45 kg

7 a 138 b 5720 kg c 11 916 kg

8 a 3975.73 kg b 37.58 kg c 27.85 kg d 18.37 kg

CHALLENGE: Teacher to check.

9 a K371.25 b K235.13 c K569.25 d K462.83 e K727.65

10 a K323 b K522.75 c K416.50 d K690.63 e K924.38

11 15.5 kg

12 K32.75 per kg

13 K334.38

14 K55 per kg

Lesson 8 – What is my money worth?

1 a 40 Australian dollars b 60 Australian dollars
c 120 Australian dollars d 100 Australian dollars
e 200 Australian dollars

2 a 895.50 baht b 1074.60 baht c 2388 baht
d 1492.50 baht e 2865.60 baht

3 a 1900 yen b 4560 yen c 2850 yen
d 14 250 yen e 6840 yen

4 a 42 euros b 15.60 euros c 28.80 euros
d 66 euros e 56.40 euros

5 a 127.50 Singapore dollars b 47.50 English pounds
c 92.50 US dollars d 140 Fiji dollars
e 722.50 South African rand f 3972.50 Philippines pesos

6 a 175.75 US dollars b 18 050 yen
c 90.25 English pounds d 1372.75 South African rand
e 190 Australian dollars f 266 Fiji dollars

7 a 800 Australian dollars b 840 Australian dollars

8 a 300 English pounds b 15 English pounds
c 330 English pounds

9 a 245 Singapore dollars b 10 Singapore dollars
c 260 Singapore dollars

10 a 11 917.50 Philippines pesos b 11 925 Philippines pesos

11 a 3468 South African rand b 3420 South African rand

12 a K540.54 b K300 c K267.86
d K50.35 e K145.33 f K250

13 K526.32

14 K117.65

15 K26.32

16 a K1282.05 b K1351.35 c K69.30

17 a K165.56 b K167.50 c K1.94

18 a Jerry b K64.10

CHALLENGE: Answers will vary.

Lesson 9 – Airfare deals

1 a K22117.15 b K10334.22 c K23760.57
d K12013.74 e K5515.49

2 Paris

3 a K21107.36 b K42134.15 c K81532.62
d K12496.40 e K57474.15 f K63168.77

4 a K4080.90 b K785.22 c K5577.38

5 a K23218.10 b K11868.27 c K26984.93
d K21213.43 e K16530.20 f K68557.02

6 a K16223.32 b K19245.43 c K32567
d K8341.35 e K29233.70 f K28696.42

7 a K4900 b K5700 c K8300 d K9700
e K7200 f K6325 g K5067 h K4150
i K5738 j K2163

8 a K3600 b K8100 c K5400 d K7200
e K6300 f K3862 g K1917 h K8366
i K7428 j K5919

9 K3080

10 K5835

11 K6675

12 K10957

13

	Full fare	30% off full fare	25% off full fare	10% off full fare
a	**K7543**	K5280.10	K5657.25	K6788.70
b	K4923	**K3446.10**	K3692.25	K4430.70
c	K5164	K3614.80	K3873	**K4647.60**
d	K3442	K2409.40	**K2581.50**	K3097.80
e	K2713	**K1899.10**	K2034.75	K2441.70
f	**K9715**	K6800.50	K7286.25	K8743.50
g	K6338	K4436.60	K4753.50	**K5704.20**
h	K5999	K4199.30	**K4499.25**	K5399.10
i	**K4289**	K3002.30	K3216.75	K3860.10
j	K8392	**K5874.40**	K6294	K7552.80

Learning Unit 3 – Tourism Comparisons

Lesson 1 - Introduction

1 A = United States of America; B = United States of America; C = England; D = France; E = Italy; F = China; G = Japan; H = Australia; I = India; J = Egypt

2 Answers will vary.

3 a Disneyland (Tokyo) b Uluru
c Disneyland (Tokyo); Great Wall of China; Eiffel Tower; Grand Canyon; Statue of Liberty; Pyramids of Giza; Tower of London; Taj Mahal; Leaning Tower of Pisa; Uluru

Lesson 2 – Price comparisons around the world

1 USA – K9.17
Australia – K8.88
England – K10.49
China – K4.30
Denmark – K15.55
France – K12.80
Hong Kong – K4.14
Indonesia – K4.63
Japan – K7.24
Norway – K21.36
Philippines – K5.33
Singapore – K7.77

2 a K12.48 b K1.40 c K5.06

3 a Norway b Hong Kong

CHALLENGE: Teacher to check.

4

	New York	London	Paris	Tokyo	Dublin
Loaf of bread	K7.94	K5.85	K8.78	K16.30	K5.85
2 kg bag potatoes	K6.06	K11.50	K10.45	K19.86	K10.87
Cup of coffee	K11.91	K12.33	K14.63	K14.42	K12.54
Bus/tube ticket	K6.35	K18.60	K5.85	K8.15	K6.27
CD (music)	K55.39	K80.04	K79.42	K56.43	K83.60

5 Answers will vary.

6 a K2.72 b K24.65 c Answers will vary.
d Tokyo, Paris, New York, London and Dublin

7 New York = K87.65; London = K128.32; Paris = K119.13; Tokyo = K115.16; Dublin = K119.13
Most expensive to least expensive: London, Paris and Dublin (the same), Tokyo, New York

8 a K59.58 b K61.65 c K38.10
d K317.68 e K48.90 f Answers will vary.

9 a US$253.45
b Papua New Guinea would come after Brazil and before India.
c In Papua New Guinea the cost of a 2GB iPod Nano is the second most expensive in the world.

10 a K183.51 b K105.82 c K58.57

11 a K41.25 b K5.36 c K150
d K8.11 e K39.47 f K33.33

12 a K200 b K115.83 c K108.75
d K535.14 e K633.32 f K324.32

13 K130.84

14 K211.11

15 K1050.70

CHALLENGE: Answers will vary.

Lesson 3 – Physical comparisons

1 a 208.4 m b 60.5 m c 23.7 m
d 229.3 m e 129.8 m f 154.7 m

2 463.3 m

3 a 65.6 feet b 328 feet c 147.6 feet
d 190.2 feet e 242.7 feet f 219.8 feet

4 a Sears Tower, CN Tower, Taipei 101
b Eiffel Tower, Chrysler Building, Eureka Tower, Burj Al Arab Hotel
c 185.2 feet

5 a 25 m b 100 m c 45.7 m
d 131.1 m e 83.8 m f 56.7 m

6 $82 \times 0.3 = 24.6$; $328 \times 0.3 = 98.4$; $150 \times 0.3 = 45$; $430 \times 0.3 = 129$; $275 \times 0.3 = 82.5$; $186 \times 0.3 = 55.8$

7 29 035 feet

8

Mountain	Height (metres)	Height (feet)
K2	8612	**28 247**
Kangchenjunga	8586	**28 162**
Lhotse	**8501**	27 883
Makalu	**8462**	27 755
Cho Oyu	8201	**26 899**

9 4509 m

10 a 49.6 miles b 74.4 miles c 21.7 miles
d 60.1 miles e 94.2 miles f 144.5 miles

11 a 104.8 km b 125.8 km c 185.5 km
d 359.7 km e 235.5 km f 462.9 km

12 $65 \times 1.6 = 104$; $78 \times 1.6 = 124.8$; $115 \times 1.6 = 184$; $223 \times 1.6 = 356.8$; $146 \times 1.6 = 233.6$; $287 \times 1.6 = 459.2$

13 698 miles

14

River	Length (km)	Length (miles)
Nile	**6710**	4160
Amazon	6452	**4000**
Yangtze	6394	**3964**
Congo	**4384**	2718
Mekong	4194	**2600**
Mississippi	3742	**2320**

15 23.87 miles

16 Hangzhou Bay Bridge is longer by 3.58 km or 2.22 miles.

17 17.19 km

Lesson 4 – Where the people go

1 a Fiji b Micronesia c Fiji

2 a about 56 000 b about 104 000 c about 50 000

3 Fiji, Papua New Guinea, Vanuatu

4 a Fiji had almost eight times the number of visitors that Papua New Guinea had.
b Vanuatu had almost three times the number of visitors that Micronesia had. (Answers will vary.)

5 a Papua New Guinea b New Caledonia
c Vanuatu d Micronesia

6 a 55 000 b 50 000 c 85 000

7 73 000

8 a Vanuatu had almost three times the number of visitors that Micronesia had.
b–c Answers will vary.

9 Graphs will vary. Teacher to check.

10 a Papua New Guinea, New Caledonia, Vanuatu
b Papua New Guinea
c 6000

11 Singapore

12 a Singapore b Indonesia c New Zealand d Australia

13 1997 to 1998. Reasons will vary.

14 a Singapore, Australia and Indonesia b Singapore
c Singapore = 1 250 000; Australia = 100 000; Indonesia = 575 000
d Answers will vary.

15 a -21% b 12% c -4%
d 28.1% e 10.9% f 15.8%

16 a–c Answers will vary.

17 Graphs will vary. Teacher to check.

18 Answers will vary.

Lesson 5 – Arrivals in Papua New Guinea

1 a 2007 b 2001 and 2002
c 24 000 d 46 000

2 a 14.7% b 33.3% c 85.7%

3 Students to draw a line graph. Teacher to check graph and predictions that are made.

4 2000, 2001, 2003, 2004, 2005, 2006 and 2007

5 a 3.2% b 5.9% c 8% d 5.1%

6 Australia – 33.1%; USA – -1.1%; Japan – -15.6%

7 a 39% b 61%

8 The pie chart drawn by students will look similar to this:

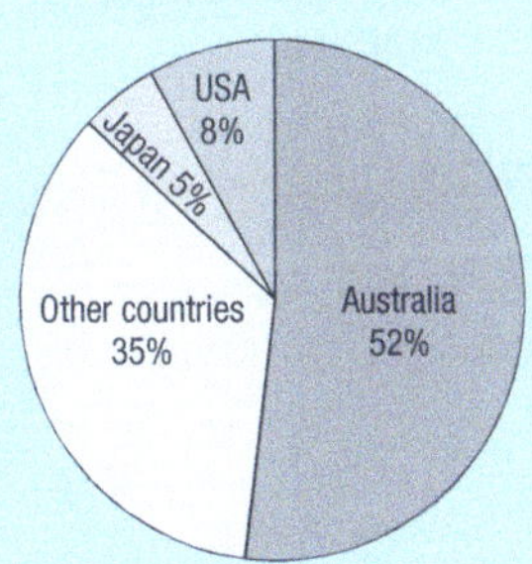

9 a 35% b 65%

10 Australia – 33.2%; USA – -22.8%; Japan – -25.6%

11 a Australia = 151.4% b Japan = 37.7% c USA = 6.3%
d Australia e USA

12 a 25.2% b 46.1% c 70.9%

CHALLENGE: Graphs will vary depending on the country chosen by students.

13 a 81 b 144 c 243 d 306 e 5670

14 a 75 b 100 c 155 d 215 e 395

15 a 17 720 b 3086 c 3689

Lesson 6 – The cost of accommodation around the world

1 a K1322.56 b K2876.72 c K3580 d K2145.60 e K1437.76

2 a K2277.72 b K1877.28 c K2653.14 d K2885.01 e K1656

3 K6888.20

4 K17095.44

5 a K561.71 b K202.43 c K242.49 d K631.03 e K425.27

6 a Auckland b Stockholm c Tokyo d Nadi

7 a K359.59 b K344.41 c K179.72 d K224.94 e K199.98

8 a K1326.57 b K1442.50 c K1050.93 d K1138.86 e K1342.50

9 a K607.20 b K984.50 c K590.04 d K363.70 e K494.87

10 a K683.32 b K323.50 c K359.96 d K865.50 e K563.18

11 a K850.25 b K665.59 c K683.22 d K654.38 e K524.40

12 a K724.14 b K1074 c K892.72 d K2733.27 e K3581.73

13 a K402.75 b K315.28 c K281.59 d K179.98 e K161.75

14 a K140.15 b K208.16 c K160.11

15 a K1205.30 b K652.89 c K893.06

16 a K1006.10 b K1251.30 c K900.93

17 a 11.7% b 24.6% c 36.8% d 35.1%

Lesson 7 – Cargo and passenger statistics

1 a 5 704 650 metric tonnes b 6 259 598 metric tonnes c 11 131 155 metric tonnes d 4 129 428 metric tonnes e 8 071 831 metric tonnes

2 a 2 048 748.3 metric tonnes b 1 907 934.3 metric tonnes c 3 605 267.3 metric tonnes

3 a Charles De Gaulle, Singapore and Los Angeles
b Los Angeles and London

4 a -0.7% b -3.3% c -1.6% d -5.9%

5 a 4.5% b 8.4% c 4% d 5.4%

6 a 1 975 704 b 3 955 791 c 2 911 294 d 1 437 786

7 a 1 945 005 b 1 821 540 c 3 659 493 d 2 103 954

8 a 22 707 852 b 35 642 364 c 29 459 904 d 41 585 685 e 37 257 073

9 a 76 565 820 b 61 475 337 c 47 798 567

10 a 10.5% b 2.6% c 8.5% d 5.7%

11 a -1.2% b -1.1%

12 51 108 393

13 a 74 274 658 b 62 546 657 c 57 352 322 d 80 005 722

CHALLENGE: Graphs will vary

Lesson 8 – Some interesting comparisons

1 China = 140.89; India = 368.41; USA = 32.57; Indonesia = 134.39; Brazil = 22.24; Pakistan = 212.92; Bangladesh = 1100.48; Australia = 2.66; Hong Kong = 6660.68; Papua New Guinea = 12.52; Singapore = 6519.81; New Zealand = 15.21; Vanuatu = 17.13; Tonga = 159.74; Monaco = 16687.18

2 From largest to smallest population density: Monaco, Hong Kong, Singapore, Bangladesh, India, Pakistan, Tonga, China, Indonesia, USA, Brazil, Vanuatu, New Zealand, Papua New Guinea, Australia

3 a China b China c Monaco

4 a Monaco b Monaco c Australia

5 a Monaco has a very small land area compared to its population.
b Australia has a very large land area compared to its population.

6 a USA, Brazil, Australia, Papua New Guinea, New Zealand, Vanuatu
b Monaco, Hong Kong, Singapore

7 a 2.67 b 12.74 c 15.58 d 299.01

CHALLENGE: Answers will vary.

8 a 906 009 b 57 463 c 66 199 d 238 381

9 a 1 566 871.17 km^2 b 1 268 816.36 km^2 c 743 214.01 km^2 d 459.95 km^2

10 a 0.7 b 4.8 c 1.8 d 5.2 e 8.7

11 a 7.1 b 8.5 c 2.6 d 5.9 e 4.1

12 a 5 b 3.5 c 4.1 d 3.9 e 6.9

13 a–b Graphs will vary depending on the scale chosen.

14 a $\frac{9}{20}$ b 45%

15 a $\frac{6}{20}$ b 30%

Learning Unit 4 – Additional Learning, Revision and Assessment

Lesson 1 – Additional learning

1 a 90° b 90° c 90° d 90°

2 a Estimates will vary.
b Angle A = 90°; Angle B = 45°; Angle C = 225°
c $\frac{90}{360}$; $\frac{45}{360}$; $\frac{225}{360}$
d $\frac{1}{4}$; $\frac{1}{8}$; $\frac{5}{8}$

3 a 40 b 16 c 8

4 a A = 60°; $\frac{60}{360}$; $\frac{1}{6}$
B = 70°; $\frac{70}{360}$; $\frac{7}{36}$
C = 140°; $\frac{140}{360}$; $\frac{7}{18}$
D = 90°; $\frac{90}{360}$; $\frac{1}{4}$
b A = 80°; $\frac{80}{360}$; $\frac{2}{9}$
B = 25°; $\frac{25}{360}$; $\frac{5}{72}$
C = 110°; $\frac{110}{360}$; $\frac{11}{36}$
D = 145°; $\frac{145}{360}$; $\frac{29}{72}$
c A = 35°; $\frac{35}{360}$; $\frac{7}{72}$
B = 120°; $\frac{120}{360}$; $\frac{1}{3}$
C = 155°; $\frac{155}{360}$; $\frac{31}{72}$
D = 50°; $\frac{50}{360}$; $\frac{5}{36}$

5 A = K24; B = K28; C = K56; D = K36

6 a 930 b 720 c 300 d 210

7 a 90°; $\frac{1}{4}$; K15 000 b 90°; $\frac{1}{4}$; K15 000
c 30°; $\frac{1}{12}$; K5000 d 45°; $\frac{1}{8}$; K7500
e 45°; $\frac{1}{8}$; K7500 f 45°; $\frac{1}{8}$; K7500
g 15°; $\frac{1}{24}$; K2500

8 athletics – 72°; soccer – 144°; basketball – 42°

9 a 4° b 3°

10 Titles of pie charts will vary but pie charts should look similar to this one:

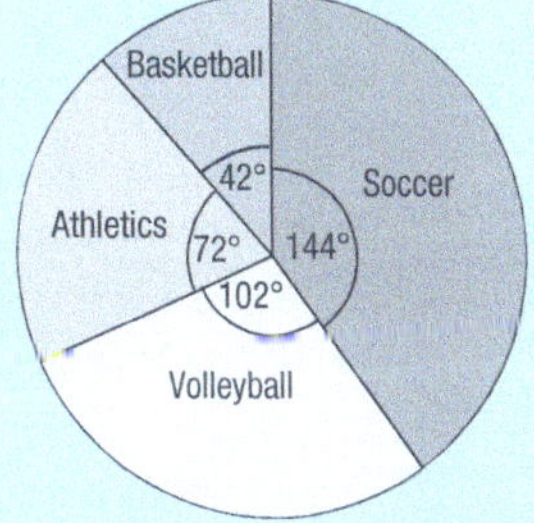

11 a 40 b 9°
c Titles of pie charts will vary but pie charts should look similar to this one:

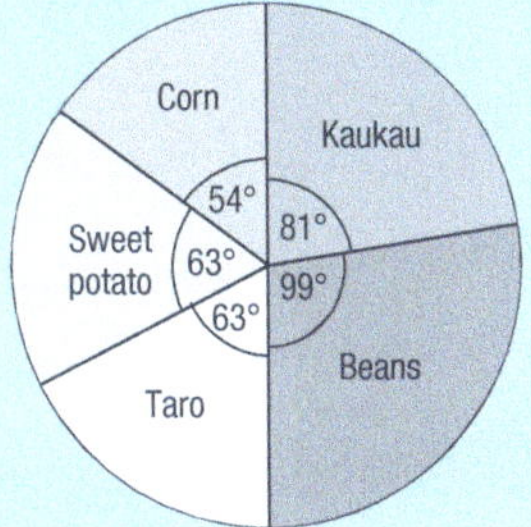

Lesson 2 – Revision

1 a 2 cm b 8 cm c 3 cm d 0.5 cm e 6 cm

2 a 10 km b 40 km c 22.5 km d 26 km e 48 km

3 a 1:200 000 b 1:500 c 1:850 000 d 1:10 000

4 a 150 km b 675 km c 560 km d 320 km

5 a 4 hours b 2 h 30 min c 4 h 36 min d 4 h 15 min

6 a K284.90 b K638 c K359.70 d K501.60
e K125.40 f K324.50

7 a K118 b K63.60 c K104 d K109.50
e K27.60 f K106.20 g K38.40 h K77.40

8 a A, C and D b Shape C

9 a 38 L b 26 L c 32 L d 20 L
e 7 L f 12 L

10 a 7.8 L b 8.4 L c 6.2 L d 6.4 L

11 a 41% b 62%

12 a south b north c north d south
e south f east g west h east

13 a Noumea b Touho c Mt. Panie d Poya

14 a 21° 35′S, 166° 10′E b 20° 35′S, 164° 20′E
c 20° 55′S, 164° 45′E d 21° 55′S, 166° 25′E

15 a 7:30 p.m. b 4:30 p.m.

16 a 2 p.m. b 3:30 p.m.

17 a 705 L b 696 L

18 a 16 hours b 7 h 30 min

19 a 4003.65 kg b 496.35 kg

20 19.35 kg

21 a K231.20 b K359.04 c K157.76 d K606.56 e K507.28

22 a $80 b $20 c $400 d $170 e $270

23 a K961.54 b K692.31 c K1153.85 d K2096.15 e K1865.38

24 a K4612.50 b K6049.50 c K5246.25 d K6845.25 e K4420.50

25 a K7150 b K9300 c K4020 d K5840 e K7990

26 a 98.4 feet b 213.2 feet c 157.4 feet
d 308.3 feet e 239.4 feet f 357.5 feet

27 a 25.5 m b 52.5 m c 71.4 m
d 37.2 m e 94.5 m f 79.8 m

28 a 25.9 m b 53.4 m c 72.6 m
d 37.8 m e 96 m f 81.1 m

29 a 41.5 miles b 78.7 miles c 54.6 miles
d 133.9 miles e 112.2 miles f 151.9 miles

30 a 152 km b 196.8 km c 115.2 km
d 252.8 km e 368 km f 315.2 km

31 a 153.2 km b 198.4 km c 116.1 km
d 254.8 km e 371 km f 317.7 km

32 a 15% b 28.3% c 11.3% d 16.5%

33 a -18.6% b -5.5% c -5% d -7.4%

34 a 69 b 86.25 c 161
d 110.4 e 124.2 f 253

35 a 76 b 100.7 c 247
d 146.3 e 204.25 f 323

36 a K1258.68 b K2936.92 c K2097.80 d K839.12

37 a K251.50 b K205.10 c K229.50
d K276.40 e K242.30 f K190.20

38 65.24

39 4 573 646

40 64 000 km^2

41 USA = 60° = 300 people; Japan = 30° = 150 people;
New Zealand = 100° = 500 people; Australia = 125° = 625 people;
Italy = 45° = 225 people

Lesson 3 – Assessment Task 1

Answers will vary (to be assessed by teacher).

Lesson 4 – Assessment Task 2

1 a $\frac{1}{5000}$ b $\frac{1}{200\,000}$ c $\frac{1}{50\,000}$ d $\frac{1}{25\,000}$

2 a K104 b K484 c K135.45 d K169.50

3 a K15 b K30 c K8.70 d K24.50

4 1870 km

5 9.5 hours

6 a 1:1 000 000 b 1:75 000 c 1:4 500 000 d 1:850 000

7 a 18.25 L b 12.78 L c 6.13 L d 9.64 L

8 a K73.50 b K88.20 c K52.92 d K129.36

9 K172.20

10 a K259.20 b K405 c K202.50 d K576.72

11 15 kg

12 a 51 b 127.50 c 91.80 d 306

13 a K50 b K120 c K250 d K37.50

14 a K473 b K412.50 c K576.40 d K322.30

15 a 37.6% b 71.2% c 78.6%

16 a -25.2% b 39.7% c 117.7% d -21.1%

17 a 94.5 b 220.5 c 141.75 d 180.6

18 a 96 b 68.8 c 222.4 d 131.2

19 a K900 b K3500 c K6200 d K4500

20 a 2 cm b 10 cm c 5 cm d 6.5 cm

21 The drawn square should be 3 cm × 3 cm

22 It means 40° north of the equator and 85° west of the prime meridian.

23 a 3855.6 kg b 644.4 kg

24 Answers will vary but bags must have a total weight of 141.5 kg.

25 a 4 a.m. b 12 noon

26 a 9 a.m. b 7 a.m.

Notes

Notes

Acknowledgments

The authors and the publisher wish to thank the following copyright holders for reproduction of their material.

Photos

Alamy/Kaehler Wolfgang, p. 80; Alamy/Photolibrary/geogphotos, p. 73; Alamy/Photolibrary/Keren Su/China Span, p. 79 bottom; AusAid Photo Library/Davis Peter, p. 126; Corbis/Lenars Charles & Josette, p. 86; iStockphoto/De Bruyne Hendrik, pp. 64 right, 118 bottom right; iStockphoto/Elisseeva Elena, p. 106; iStockphoto/Gertenbach Paul, p. 120; iStockphoto/Hakim Hazlan Abdul, p. 118 top right; iStockphoto/Kunz Matt, p. 108; iStockphoto/Mette Holger, p. 119; iStockphoto/Noskowski Maciej, pp. 64 left, 104, 128; iStockphoto/Quevedo Fle Joan, p. 130; Photolibrary/Peluso Tammy, p. 71; Photolibrary/Sassoon Sybil, p. 77; Photolibrary/Thibaut Nicolas, p. 79 top; Sawczak Irene, pp. 15, 19, 36, 81, 83, 85, 87, 110, 139; Shutterstock/Bratslavsky Natalia, p. 118 top left; Shutterstock/Cumming Robert, p. 105; Shutterstock/Shah Vishal, p. 118 bottom left.

Graphs

'Weight-for-age percentiles' graphs on page 9 are reproduced by kind permission of Steve Halls, http://www/halls.md/.
'Who is infected? September 2005' graph on page 26 and 'New and cumulative confirmed cases – December 2005' graph on page 30 are from *HIV & AIDS & STI Basic Training Resource Book*, p. 12, National AIDS Council of Papua New Guinea.

Every effort has been made to trace the original source of copyright material contained in this book. The publisher will be pleased to hear from copyright holders to rectify any errors or omissions.